Cardiac Electrophysiology 2

An ADVANCED Visual Guide for Nurses, Techs, and Fellows

VIDEO ACCESS

To access the videos mentioned in this book, please visit: https://bit.ly/purvesadvanced

Cardiac Electrophysiology 2

An ADVANCED Visual Guide for Nurses, Techs, and Fellows

**Paul D. Purves, George J. Klein, Peter Leong-Sit,
Raymond Yee, Allan C. Skanes, Lorne J. Gula, Jaimie Manlucu**

cardiotext

Minneapolis, Minnesota

Contents

continues

Contents *continued*

About the Authors

Paul D. Purves, BSc, RCVT, CEPS

Senior Electrophysiology Technologist
Cardiac Investigation Unit
London Health Sciences Centre
London, Ontario, Canada

George J. Klein, MD, FRCP(C)

Professor of Medicine
Division of Cardiology
Western University
London, Ontario, Canada

Peter Leong-Sit, MD, FRCP(C)

Assistant Professor of Medicine
Division of Cardiology
Western University
London, Ontario, Canada

Raymond Yee, MD, FRCP(C)

Professor of Medicine
Director, Arrhythmia Service
Division of Cardiology
Western University
London, Ontario, Canada

Allan C. Skanes, MD, FRCP(C)

Professor of Medicine
Division of Cardiology
Western University
London, Ontario, Canada

Lorne J. Gula, MD, FRCP(C)

Associate Professor of Medicine
Division of Cardiology
Western University
London, Ontario, Canada

Jaimie Manlucu, MD, FRCP(C)

Assistant Professor of Medicine
Division of Cardiology
Western University
London, Ontario, Canada

Foreword

The process of moving beyond basic EP concepts and applying them to live cases is the ultimate goal of any member of an electrophysiology team. Learning how to do so can be quite challenging, and is best achieved in a supportive, educational environment.

This unique, interactive guide functions to build on the basic concepts covered in the previous edition by helping the reader work through commonly encountered scenarios in the lab.

As we do with our trainees, this book emphasizes a systematic approach to the interpretation of EP tracings, and the importance of making an effort to understand each event as it occurs. Although induction of tachycardia and the use of relevant diagnostic maneuvers are often very informative, spontaneous events that often occur during the intervening periods can also be very informative. Many interesting displays of basic physiology can take place unprovoked, which can provide clues to the patient's underlying substrate and potential mechanisms of their clinical tachycardia. Much can be learned from these spontaneous events. They are often more challenging to understand than a tachycardia's response to standard entrainment maneuvers.

This carefully selected collection of cases is another collaborative effort by our EP team, coordinated by Paul Purves, our lead EP technologist. Paul has been an integral part of our EP team for over 15 years. His experience and breadth of knowledge has made him invaluable to our service. He has significantly contributed to the education of innumerable technologists, nurses and EP fellows over the years. He is a patient and dynamic teacher whose enthusiasm for the craft is reflected in this book.

The journey to enlightenment in EP is an exciting and at times, challenging one. We sincerely hope you enjoy this new installation, and hope that you find it helpful as you move on to the next level.

Jaimie Manlucu
Western University
London, Ontario, Canada

Preface

Cardiac Electrophysiology 2: *An Advanced Visual Guide for Nurses, Techs, and Fellows*...is just that—a more advanced examination of electrophysiology as compared to the first edition. Although it is still intended for allied health professionals, technologists, and EP fellows, it builds upon the basic concepts highlighted in the previous book. Therefore, a solid understanding of "basic" electrophysiology is assumed.

The tracings presented are very much "every-day" observations. The cases are designed to guide you towards a more in-depth understanding of cardiac electrophysiology and encourage you to pursue further readings on the concepts introduced.

This book is designed as a "workbook." Each case begins with a tracing paired with some background clinical information and a question or challenge. Turning the page reveals the same tracing with arrows and other clues that highlight the key aspects of the tracing. The final page of each case contains a full explanation of each of the key learning points in the tracing.

Enjoy!

Acknowledgements

This second book would not have happened without the ongoing collaborative environment I work in. This environment has allowed all of us to learn "on the job" and our ongoing invigorating Friday morning arrhythmia rounds continues to be a fertile feeding ground for contribution and learning.

A special thanks, once again, to the great nurses and technologists with whom I have had the pleasure to work and collaborate with and who have had an unspoken impact on this effort:

Marilyn Braney, RN
Nicole Campbell, RCVT
Barb Prudhomme, RN
Jane Schieman, RN
Laura Spadini, RCVT
M.J. Vanstrien, RN

Thanks to my family—my wife Judy and daughters Mandy and Kelly—for their ongoing support and encouragement.

Cover artwork and design by Mandy Trainor and Caitlin Altobell.

—*Paul D. Purves*

Glossary and Abbreviations

A   atrial

Accessory pathway (AP)   An additional electrical connection (other than the AV node) between the atria and ventricles. Accessory pathways may conduct:

1) Antegrade only...generates a delta wave
2) Retrograde only... no delta wave
3) Bidirectional... generates a delta wave

Action potential   The waveform generated by a cell's depolarization.

AEGM   atrial electrogram

AF   *see* Atrial fibrillation

A-H interval   Transit time through the AV node.

Anisotropy   The concept that conduction velocity along a cardiac muscle fiber is determined by angle of initial depolarization.

Antidromic   Denoting a wave of depolarization that is traveling retrogradely through the AV node. This term is usually used when describing the direction of the wavefront of an atrioventricular reentrant circuit. In this scenario, the wave of depolarization travels down the accessory pathway and back up the AV node. It can also denote a wave of depolarization that is traveling in the opposite direction to the predominate wave. For instance, a clockwise wave of depolarization introduced into a counterclockwise atrial flutter, therefore called an antidromic wavefront.

AP   *see* Accessory pathway.

Ashman phenomenon   When a relatively long cycle (R-R) is followed by a relatively short R-R, the QRS associated with the short R-R often has right bundle branch block (RBBB) morphology. Often referred to a "long-shorting" the bundle branches. The right bundle branch has had insufficient time to adapt its ERP to a sudden change in heart rate. It therefore blocks.

AT   *see* Atrial tachycardia

Atrial fibrillation (AF)   A disorganized atrial rhythm associated with irregularly irregular ventricular response.

Atrial tachycardia   It may originate in either the left or right atrium.

Atropine   A parasympatholytic used to increase heart rate.

AV   atrioventricular

A-V   Referring to the total transit time from the atria to the ventricles.

AVCS   Refers to the atrioventricular conduction system.

AVNRT   AV nodal reentrant tachycardia

AVRT   atrioventricular reentrant tachycardia

Bipolar   Refers to the use of two poles (a positive and negative electrode) usually in close proximity to one another. However, a unipolar lead is just a *very* widely spaced bipole.

GLOSSARY AND ABBREVIATIONS

Bundle of His   Anatomic structure connecting the AV node to the bundle branches. It is mostly a ventricular structure.

Carto   A 3D mapping system from Biosense Webster.

CFEs   Complex fractionated electrograms usually associated with the left atrium during atrial fibrillation.

Chevron   A pattern of activation where CS 1-2 activation is as early as CS 9-10 act while the remaining poles are later.

Concealed   An unseen penetration of a wave of depolarization into a structure that subsequently affects the conduction properties of that structure.

Contralateral   "On the other side." Often used as "contralateral bundle branch block." It refers to a LBBB pattern during a right-sided tachycardia or a RBBB during a left-sided tachycardia and the effect of that bundle branch block on the tachycardia.

CS   Coronary sinus

CTI   Cavotricuspid isthmus

Decapolar   10-pole catheter. Often the CS catheter.

Decrement   The ability of the AV node to slow conduction of a wave of depolarization.

Delta wave   A slurred, initial deflection of a widened QRS that represents early conduction directly into a ventricle via an accessory pathway.

Depolarization   The process by which a cell changes on the inside from electrically negative to electrically positive.

Digitization   The conversion of an analog signal to a digital signal.

Duo-deca   A 20-pole catheter (2 × 10 poles).

ECG   Electrocardiogram

Echo   An unexpected return of a wave of depolarization to the chamber where the wave originated.

EGM   Electrogram

Entrainment   A pacing maneuver where we pace slightly faster than the tachycardia cycle length, temporarily speeding up the tachycardia rate, then stopping pacing which allows the circuit to return to its initial cycle length. The main purpose is to determine how close to the tachycardia circuit the pacing catheter is.

EP   Electrophysiology

ERP   Effective refractory period. By definition it is the longest S1S2 that fails to conduct. Most of the time we are looking for the ERP of the AV node or an accessory pathway. With a very tight S1S2 (600, 200), you may reach the ERP of the myocardium itself (seen as no capture of the S2).

Escape rhythm   A heart rhythm initiated by a focus outside the SA node.

Far-field   Cardiac electrograms that seem to be recorded either from a distance or due to poor contact. Usually smooth, low frequency electrograms.

FP AV nodal fast pathway, usually referring to the antegrade pathway.

FRP Functional refractory period. Examples: 1) Referring to the AV node, it is the shortest output (H1-H2) achievable in response to any S1S2 input, meaning that no decrement has yet to occur; 2) Referring to the ventricle, it is the shortest V1-V2 achievable from any S1S2 input, meaning that no latency has yet to occur.

Gap An unexpected resumption of conduction after block has occurred, usually observed during evaluation of AV nodal function.

His The tissue connecting the distal AV node with the proximal bundle branches.

His-Purkinje system The specialized conduction system within the ventricles that includes the His bundle, left and right bundle branches, and the Purkinje network.

HRA High right atrium

H-V interval Transit time from the His bundle to the ventricle.

Infrahisian Below the His.

Ipsilateral "On the same side." Often used as "ipsilateral bundle branch block." It refers to a RBBB pattern during a right-sided tachycardia, or a LBBB during a left-sided tachycardia and the effect of that bundle branch block on the tachycardia.

Isoproterenol (Isuprel) A sympathomimetic. Used to increase heart rate. Used in the EP lab to alter autonomic tone in the AV node.

Jump A sudden lengthening of the V-A time or A-V time of 50 ms or more with a tightening of the S1S2 coupling interval of 10 ms. A sudden, abrupt prolongation of the A-H interval.

Junctional Beats that originate from the junctional region below the AV node.

LA Left atrial

LAA Left atrial appendage

LAO Left anterior oblique

Lasso A duo-decapolar catheter. The trademarked name for a Circular Mapping Catheter from Biosense Webster.

Latency The time period between delivery of the pacing stimulus and the local capture/depolarization of the myocardium.

LBBB Left bundle branch block

LCP Lower common pathway. Most often referred to when discussing AVNRT.

LRA Low right atrium

LV Left ventricle/ventricular

Mobitz type I Second-degree AV block. Also known as Wenkebach. A prolongation of the PR interval until a P wave is not conducted.

Mobitz type II Second-degree AV block. Characterized by a constant PR interval with intermittent unexpected non-conducted sinus P waves.

MRA Mid-right atrium

ms millisecond; msec

msec. millisecond; ms

NC No capture. Implying that the S1S2 coupling interval has reached the ERP of the myocardium itself.

Near-field Cardiac electrograms that are recorded directly under the catheter electrodes. Implies excellent contact with the myocardium. Usually sharp and high frequency.

Orthodromic Denoting a wave of depolarization that is traveling antegradely through the AV node. Also a wave of depolarization traveling in the same direction as the predominant wavefront.

Overdrive suppression When another focus within the heart spontaneously depolarizes faster than the SA node, the SA node is effectively put to sleep. Since the SA node drives a heart rate of 70 beats per minute and the junction drives the heart rate at 40 beats per minute, the SA node "overdrive suppresses" the junction. Similarly, a junctional escape rhythm of 40 beats per minute will overdrive suppress a ventricular escape rhythm of 20 beats per minute.

PAC Premature atrial contraction

P-A interval Transatrial conduction time, measured from the onset of the surface P wave to the A wave on the His catheter.

Parahisian pacing Refers to pacing beside or close to the bundle of His (para = beside or next to). A pacing maneuver used to determine the presence or absence of an anteroseptal accessory pathway.

Paroxysmal Usually used as a descriptor of atrial fibrillation that lasts only a few hours. May be used with any tachycardia lasting only a few hours.

Peel back The concept where a bundle branch adapts to increased heart rate by shortening its refractory period and increasing its conduction velocity. This results in the spontaneous normalization of a bundle branch block.

Persistent Usually referring to atrial fibrillation lasting days or longer.

Posteroseptal An accessory pathway that connects the right atrium with the left ventricle. Usually located near the CS OS.

PPI Postpacing interval. A measurement made following entrainment of a tachycardia circuit.

Preexcitation Early ventricular activation via an accessory pathway. Ventricular activation earlier than that provided through the AV node.

PR interval The timing between the onset of the P wave and the onset of the QRS on the surface ECG.

PS A posteroseptal accessory pathway.

PV Pulmonary vein

PVC Premature ventricular contraction

P wave Atrial depolarization on the ECG.

QRS Ventricular depolarization on the surface ECG.

Radiofrequency   The type of energy delivered to ablate cardiac tissue. The frequency is 350–500 kHz.

RAO   Right anterior oblique

RBBB   Right bundle branch block

Refractory   Denoting a time when the tissue is unresponsive to electrical stimulation.

RV   Right ventricle/ventricular

RVA   Right ventricular apex

S1   Stimulus 1: the drive train of 8 beats.

S2   Stimulus 2: the first extra-stimulus after the drive train.

S3   Stimulus 3: the second extra-stimulus after the drive train.

SA   Sinoatrial node

SP   AV nodal slow pathway

Spontaneous normalization   The ability of either bundle branch to adapt its conduction properties in response to a change in heart rate and thus transition from a LBBB or RBBB pattern to a normal QRS morphology.

Stim-A time   From the stimulation spike to the earliest atrial activation.

SVT   Supraventricular tachycardia

Tachycardia   Rapid heart rate. By definition, greater than 100 beats per minute.

TCL   Tachycardia cycle length

Transseptal conduction   Refers to a wave of depolarization conducting across the interventricular septum. It can be from RV to LV or LV to RV. Such myocardial conduction is much slower than conduction through the specialized bundle branch system.

T wave   Ventricular repolarization on the surface ECG.

Unipolar   Refers to the use of a single pole (by convention, the positive electrode) in contact with the area of interest and a distant second electrode (by convention, the negative electrode). In essence a *very* widely spaced bipole.

VA   Ventriculoatrial

V-A   Referring to the total transit time from the ventricles to the atria.

VEGM   Ventricular electrogram

VERP   Ventricular effective refractory period. Demonstrated as "no capture" following a tight S1S2 coupling interval, e.g., 600,200.

VF   Ventricular fibrillation

VT   Ventricular tachycardia

Wenkebach   The pacing rate at which a loss of 1:1 conduction between the atria and the ventricles (or between the ventricle and the atria) occurs.

Wobble   A rather colloquial term for variability in the cycle length of a tachycardia.

WPW   Wolff-Parkinson-White

Notes

Cardiac Electrophysiology 2

CHAPTER 1   PHYSIOLOGY

1.1 Common Pitfalls

This figure nicely illustrates the importance of proper catheter positioning required for accurate baseline measurements. Identify the unexpected potential in the displayed channels.

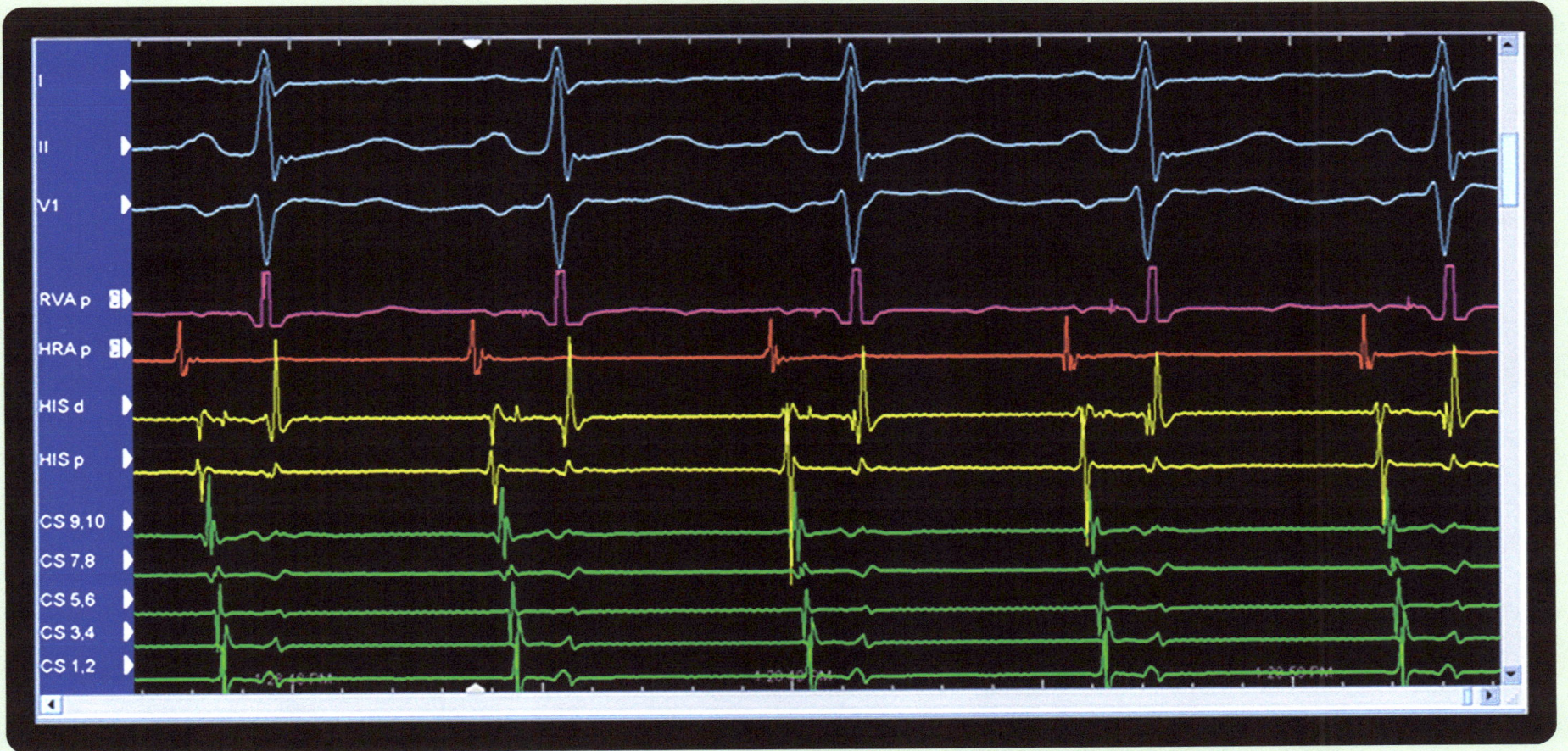

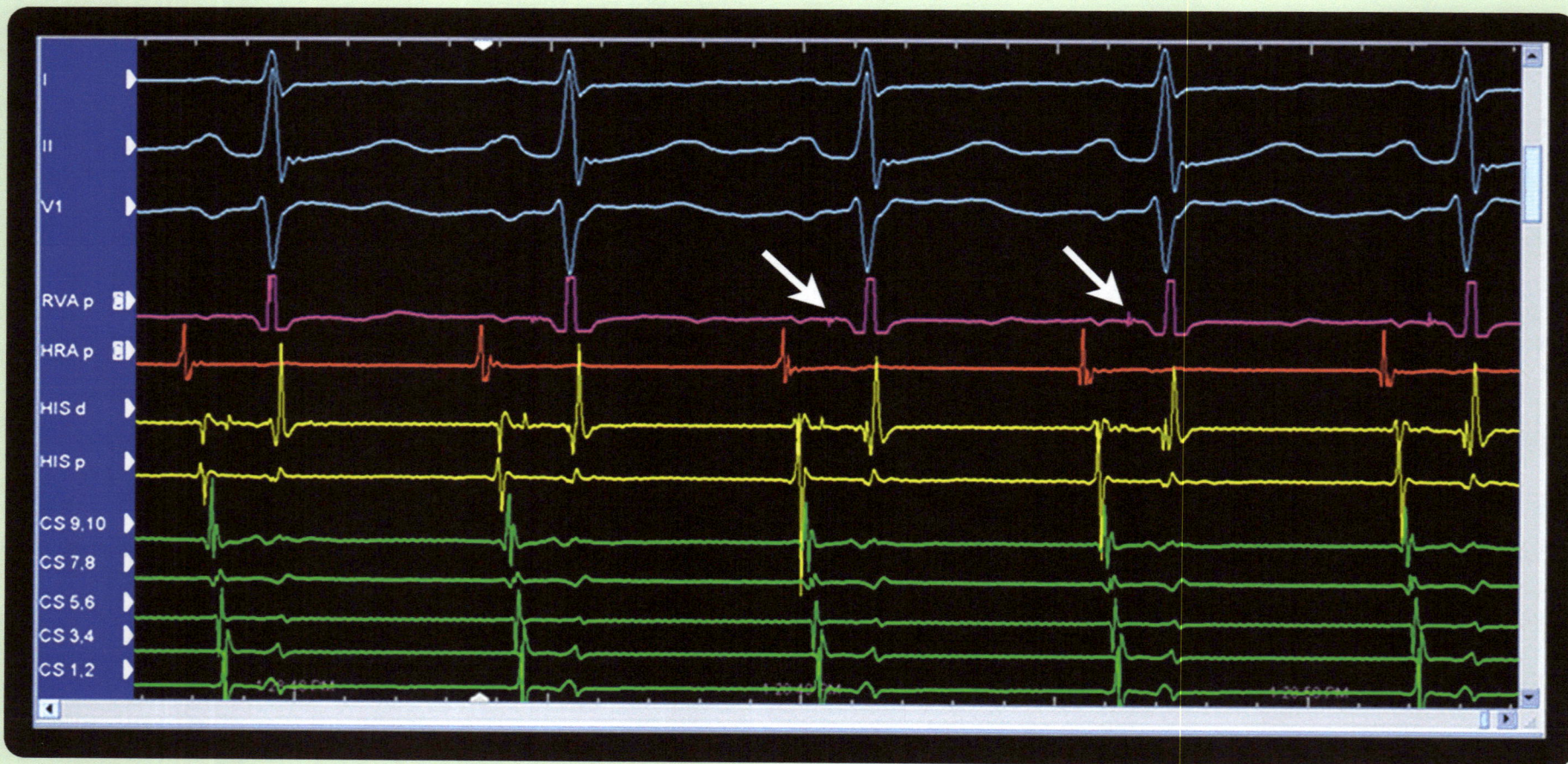

Discussion

In this example, the RV apical (RVA) catheter is positioned at the basal septum, where it can often record a right bundle potential. This should not be confused with a His potential.

Mistaking the right bundle potential for a His electrogram would underestimate the true H-V interval, leading to the erroneous conclusion that an accessory pathway (AP) is present. A normal H-V interval is approximately 35 to 45 ms.

In this particular case, mistaking the right bundle potential for a His electrogram would lead to an H-V of 10 ms, falsely suggesting that the patient is preexcited. This is clearly not the case since there is no manifest preexcitation on the surface ECG.

In this tracing, when the H-V is properly measured using the His catheter electrograms, it is within the normal range.

1.2 Heart Block

To review, there are 3 levels of heart block (or AV nodal block). Each level represents progressively more distal conduction system disease.

First-degree AV block represents AV nodal disease and manifests as a long P-R interval (greater than 200 ms) on an ECG.

Second-degree Mobitz Type I AV block (also known as "Wenkebach") represents slightly more advanced AV nodal disease, and manifests as a progressive prolonging of the P-R interval until a P wave does not conduct.

Second-degree Mobitz Type II AV block represents conduction system disease that is distal to the AV node. This manifests as a constant P-R interval with intermittent unexpected nonconducted sinus P waves.

Third-degree AV block (also known as "complete heart block") represents severe distal conduction system disease. This manifests as sinus P waves at a rate faster than the ventricular rate, A-V dissociation, and a regular ventricular response. The ventricular escape rhythm can originate anywhere from the junction to the ventricles.

Where is the level of block in this example?

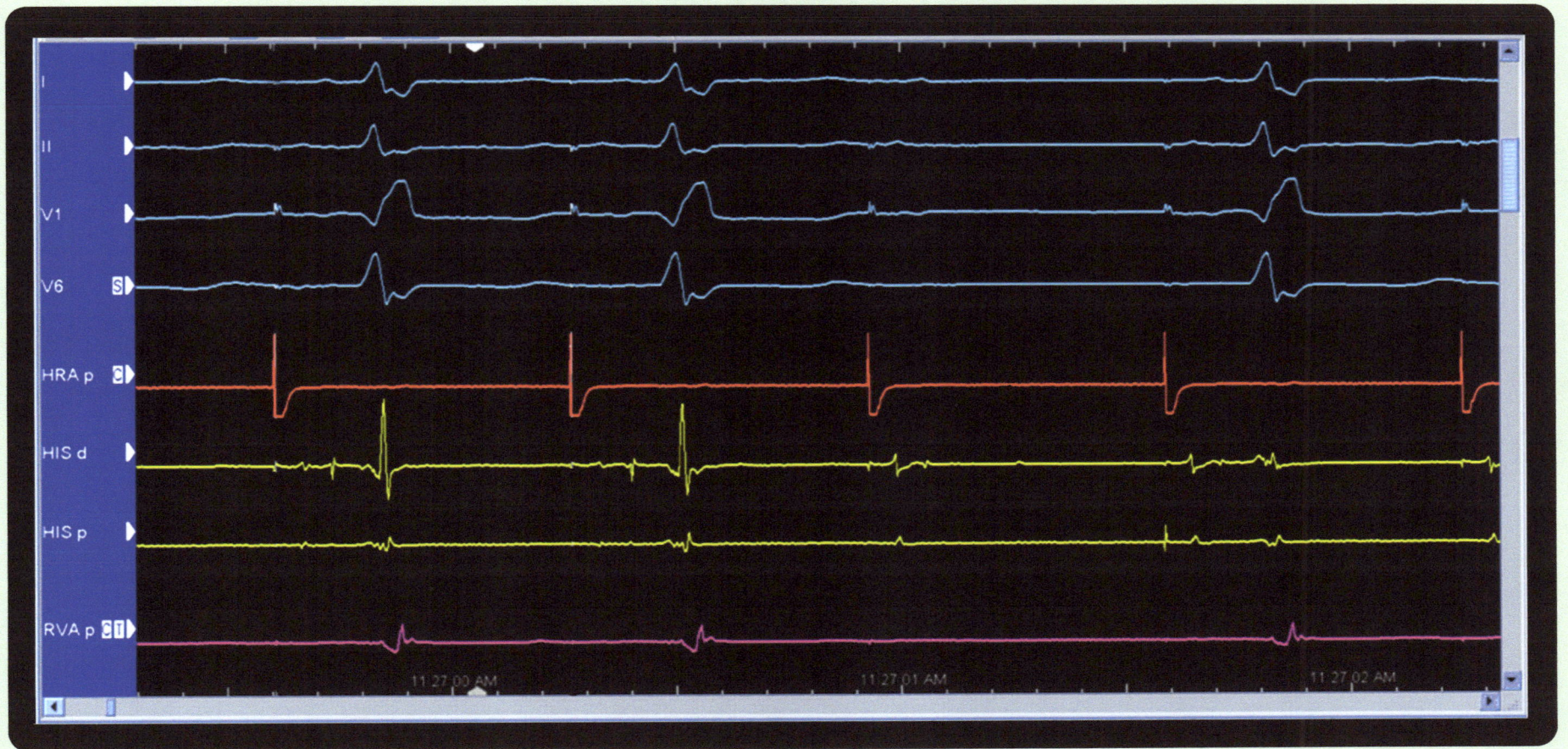

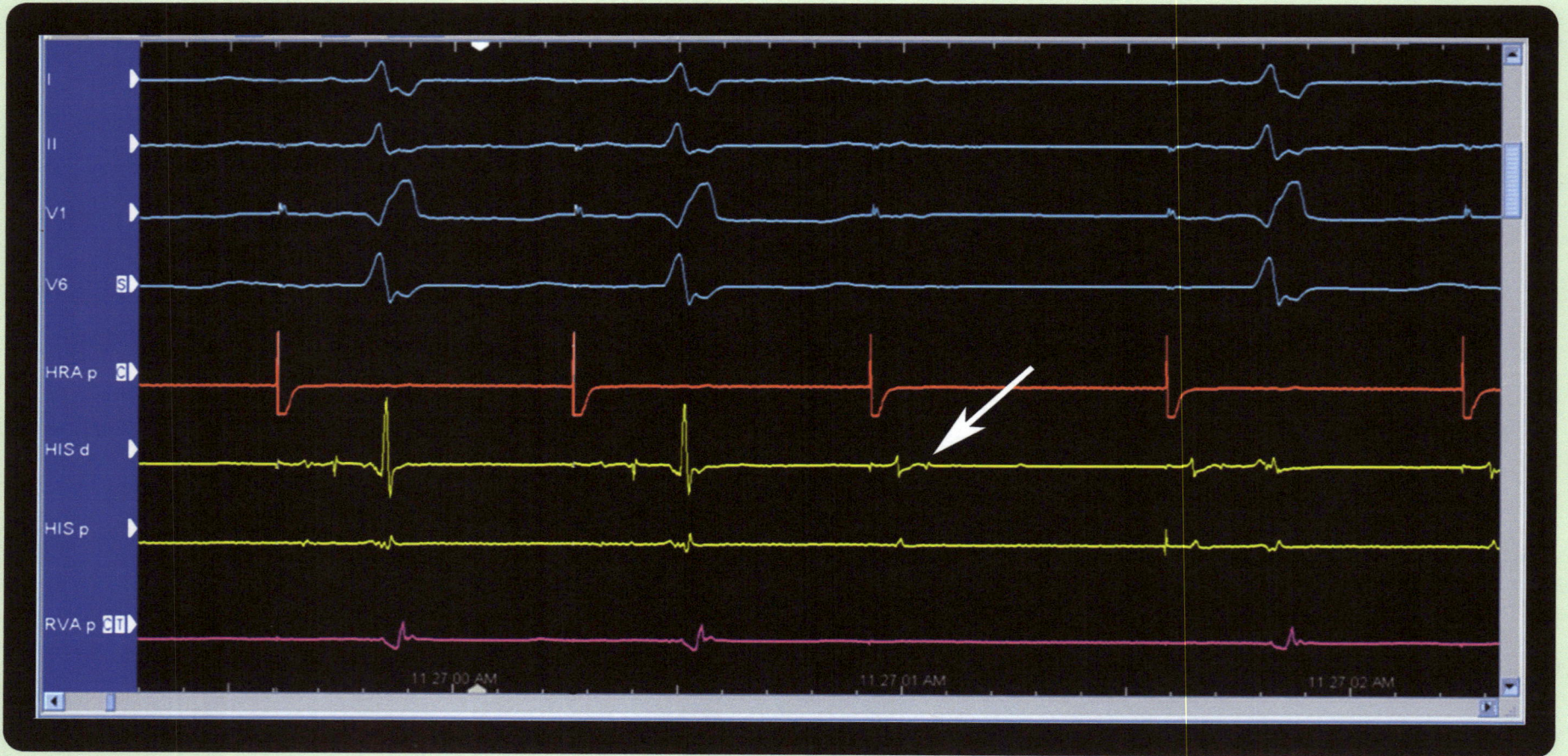

Discussion

This figure demonstrates the presence of Mobitz Type II second-degree heart block. As mentioned, this is usually indicative of distal conduction system disease.

The intracardiac electrograms show regular atrial pacing with intermittent block to the ventricles while displaying a His electrogram (*white arrow*). Lack of conduction *below* the His is indicative of the presence of conduction disease *below* the AV node.

This makes A-V conduction unreliable and often progresses unpredictably to complete heart block. Therefore, these patients are usually referred for pacemaker implantation.

1.3 Proximal Delay

In this example, we are doing ventricular extra-stimulus pacing to evaluate the retrograde Effective Refractory Period (ERP) of the AV node. The S1S2 coupling interval was 600/240 ms. But, what input did the AV node actually receive?

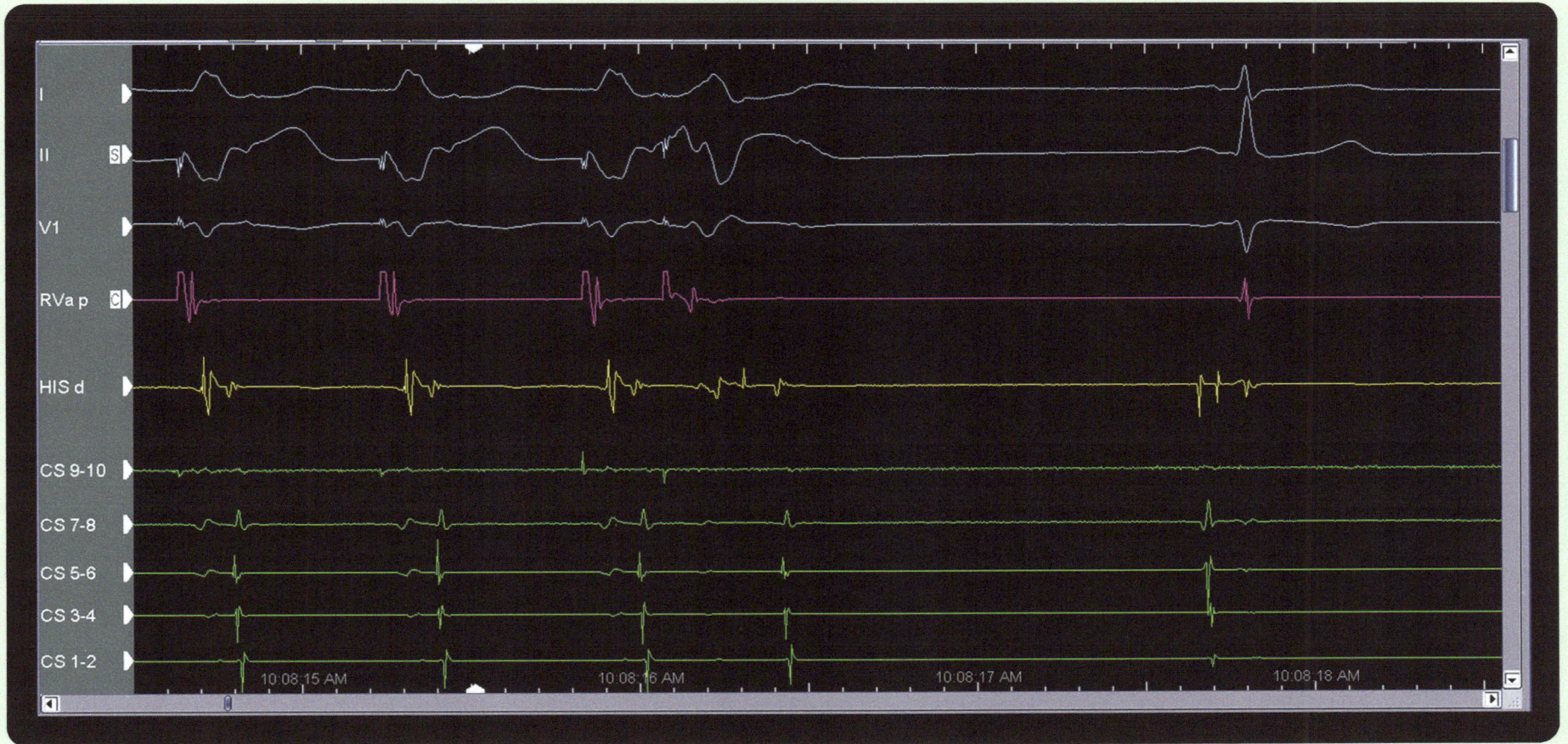

Figure 1

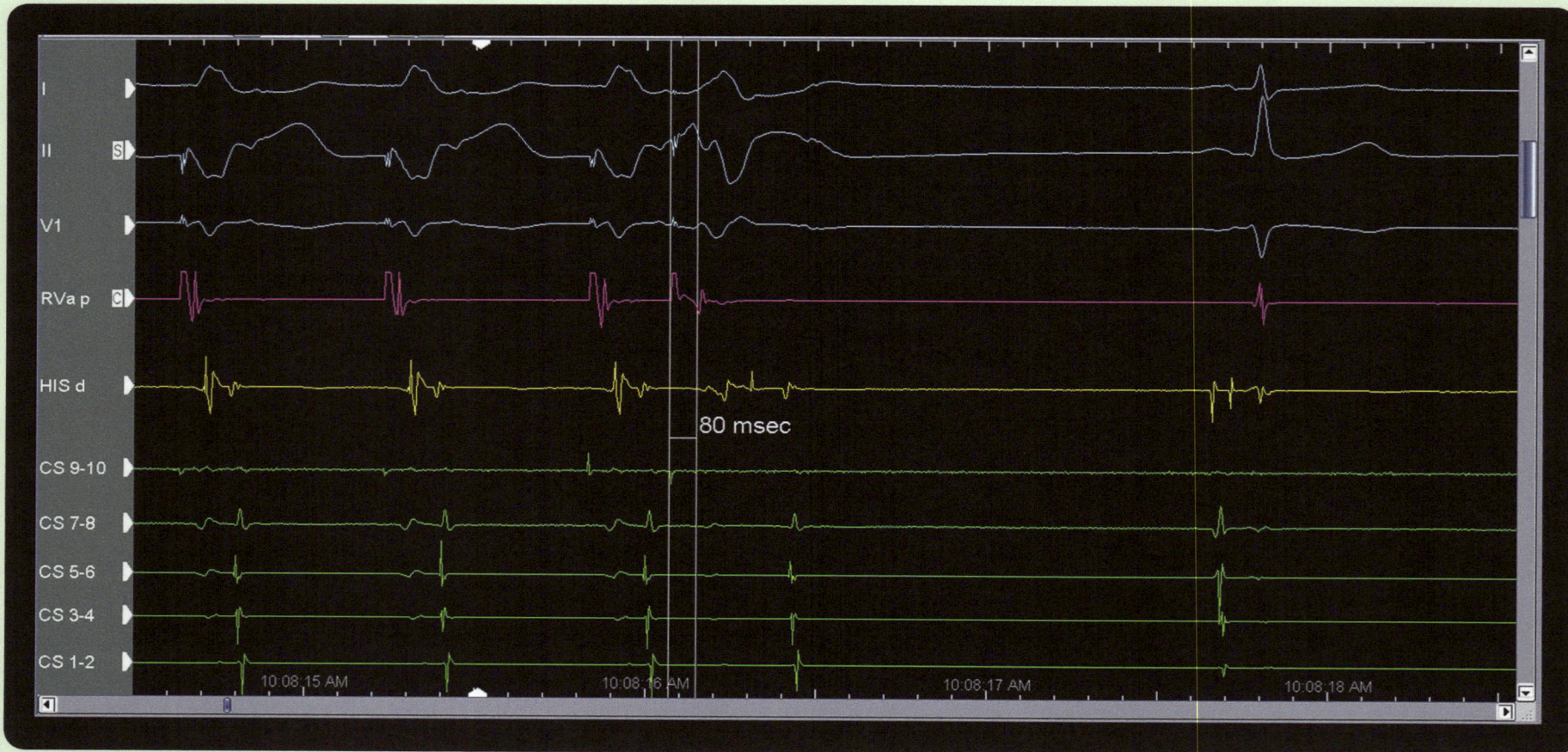

Figure 2

Discussion

Figure 2 uses calipers to highlight 80 ms of latency (local myocardial conduction delay).

continues

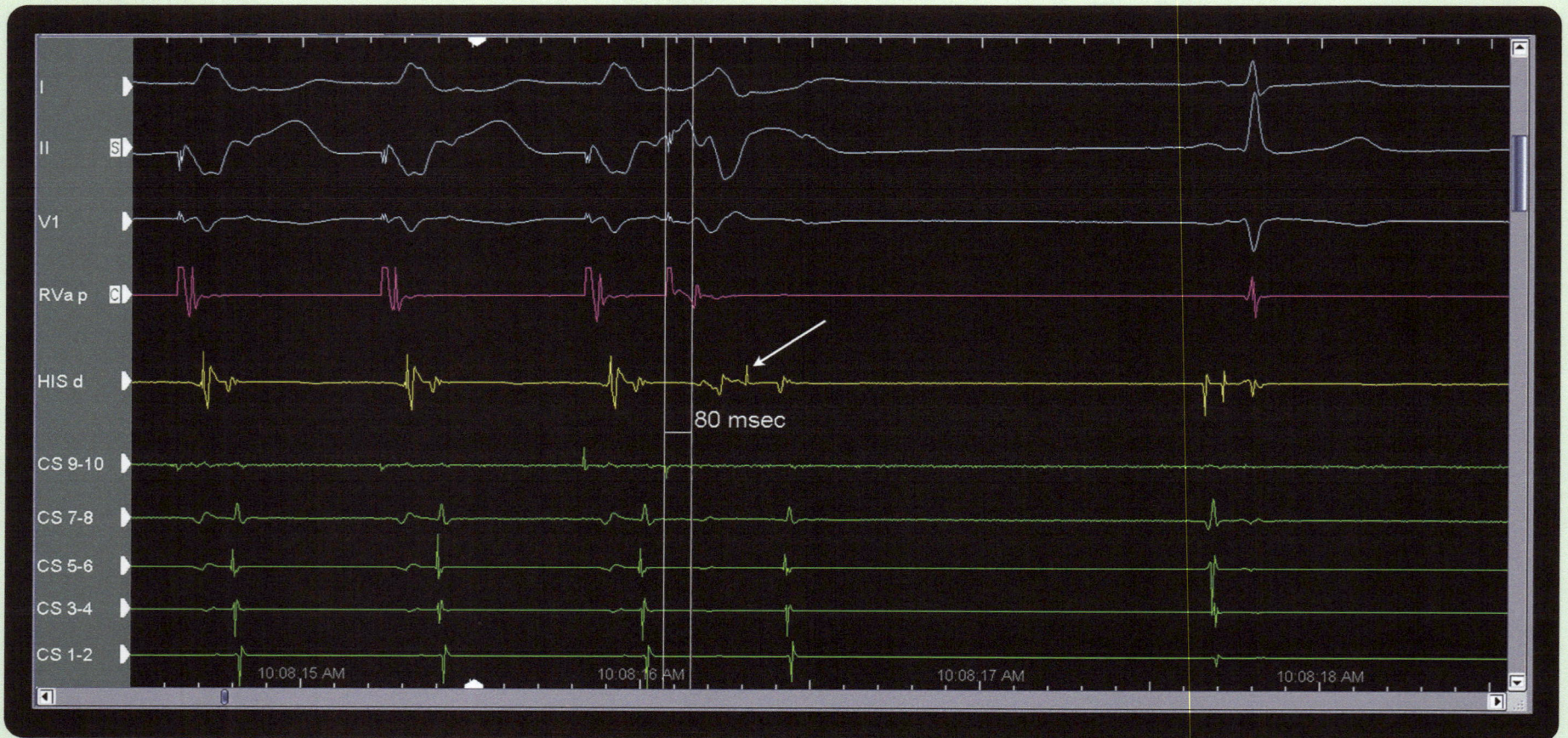

Figure 3

14

Figure 3 identifies a retrograde His (*white arrow*). The presence of a retrograde His suggests that we have reached the retrograde ERP of the right bundle. Therefore, the wave of depolarization from the S2 must travel from the RVA through the septum and retrograde up the left bundle to the His. This transseptal conduction created a <120 ms delay, thus separating the His electrogram from the local ventricular electrogram.

continues

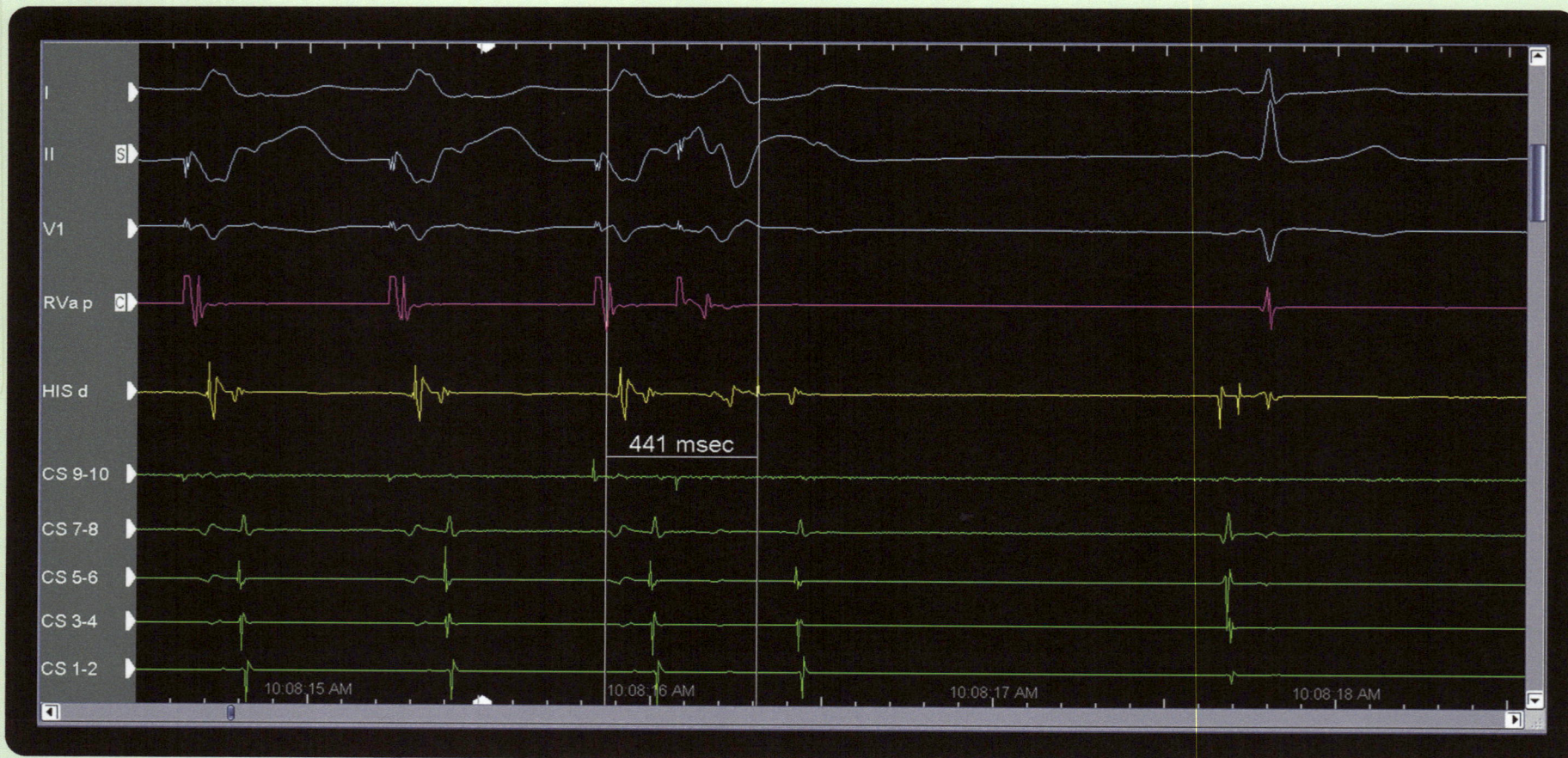

Figure 4

16

Figure 4 shows calipers indicating what input the AV node actually received, taking into account the latency and the block in the right bundle. Although the delivered S1S2 coupling interval was 600/240 ms, the AV node actually received an input of 600/441 ms (i.e., 240 ms + 80 ms of latency + 121 ms of transseptal conduction time = 441 ms).

These proximal conduction delays suggest that the ERP of the myocardium will likely be reached before the ERP of the AV node.

1.4 Aberrancy

Both of these tracings demonstrate AV nodal reentrant tachycardia (AVNRT) with aberrant conduction. Discussion of AVNRT will be covered in a subsequent chapter. These figures focus on the physiology of aberrant conduction.

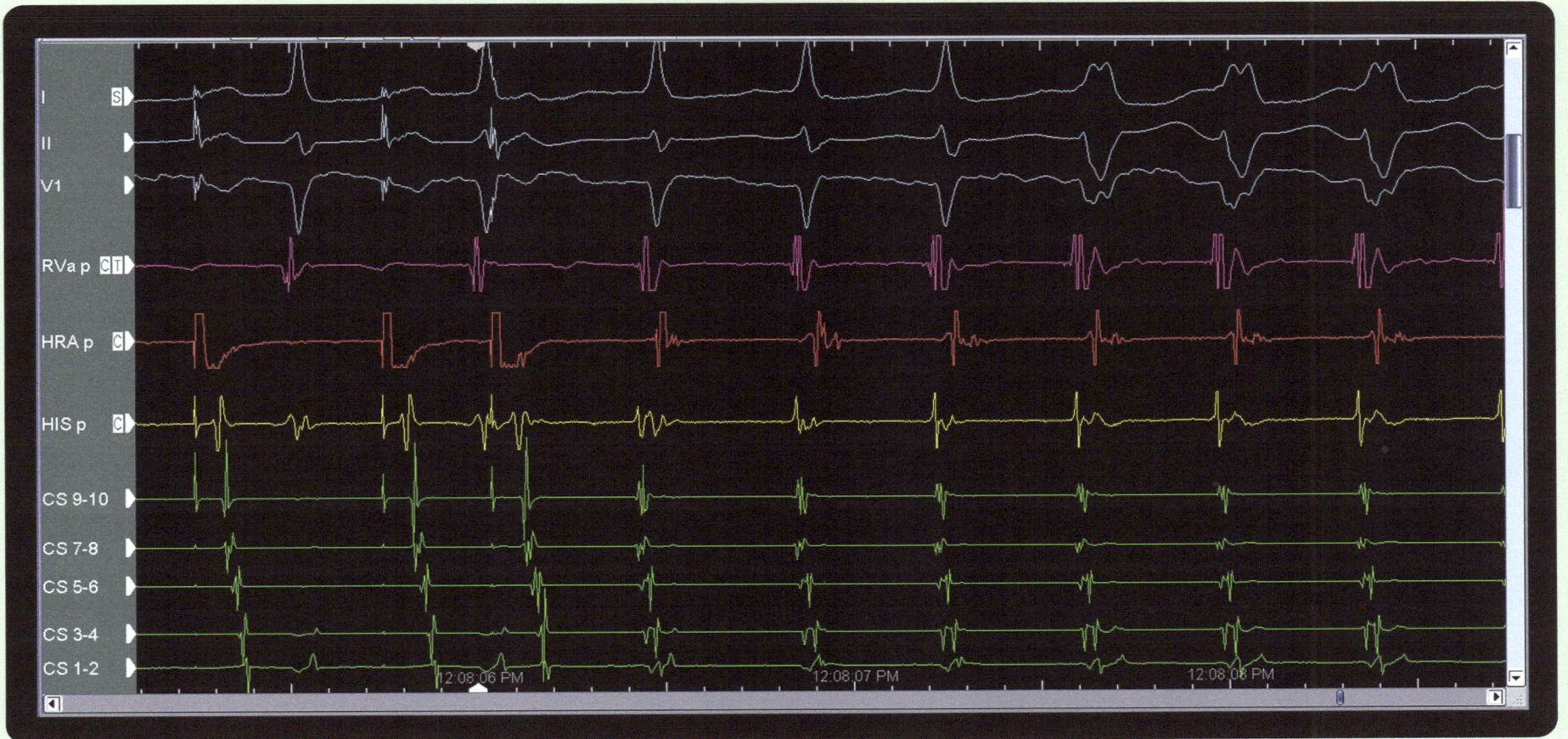

Figure 1

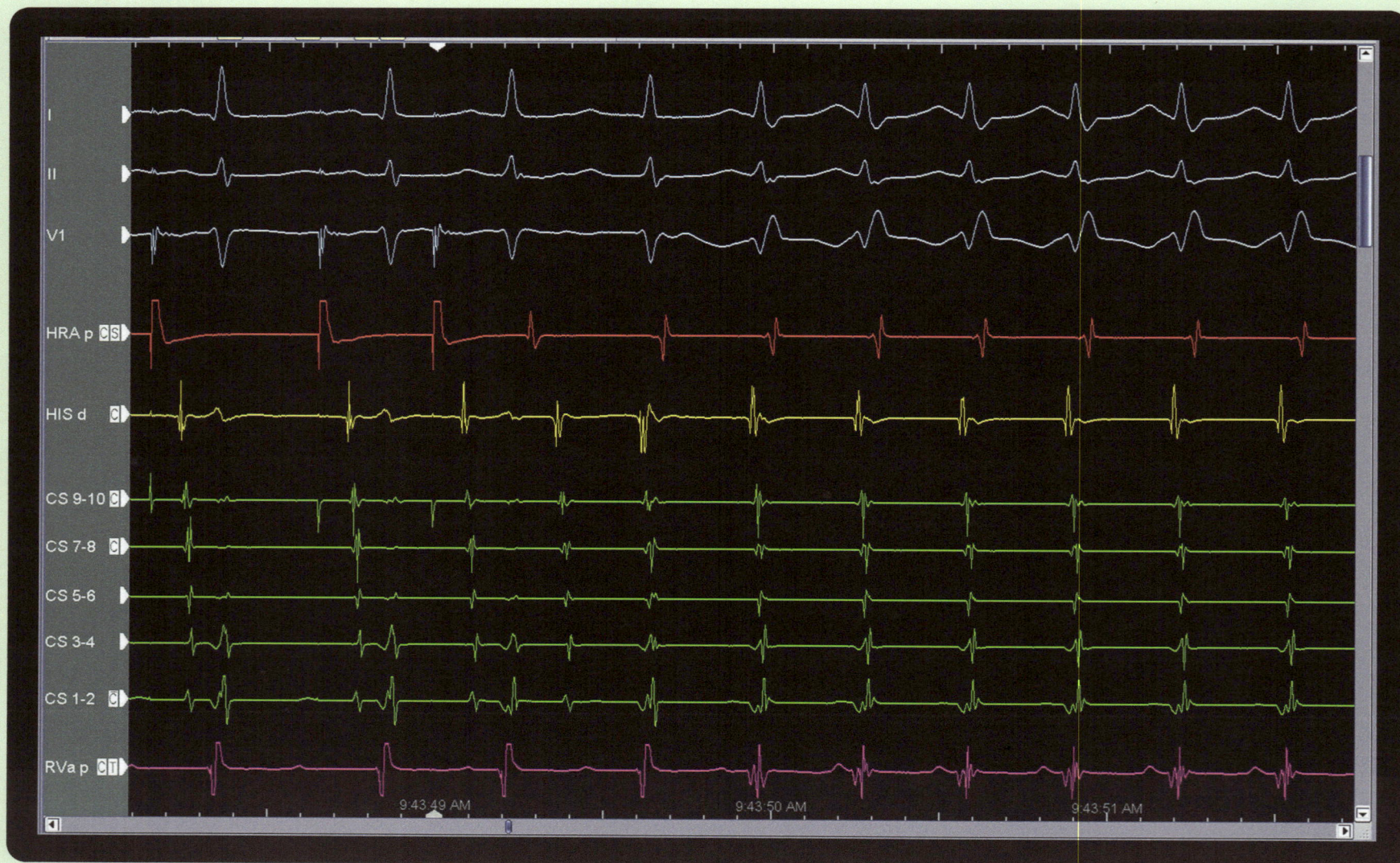

Figure 2

Discussion

In both tracings, AVNRT is initiated with atrial extra-stimulus pacing. We see development of a typical left bundle branch block (LBBB) in **Figure 1**, and a typical right bundle branch block (RBBB) in **Figure 2**.

These bundle branch blocks develop due to a sudden shortening of the tachycardia cycle length, which catches one of the bundle branches (most commonly the right bundle) still in its refractory period. This is commonly referred to as a "rate-related" bundle branch block or Phase III block. This is a functional block and is a normal phenomenon.

The bundle branch block usually spontaneously resolves due to a phenomenon referred to as "peeling back of refractoriness," a process by which the refractory period of the bundle shortens and conduction velocity is enhanced in response to increased heart rate.

See Chapter 1.5 for further clarification.

1.5 Normalization

Previously in Section 4, we discussed the reasons for aberrant conduction. However, aberrant conduction may spontaneously normalize. Please consider the reasons why this may occur.

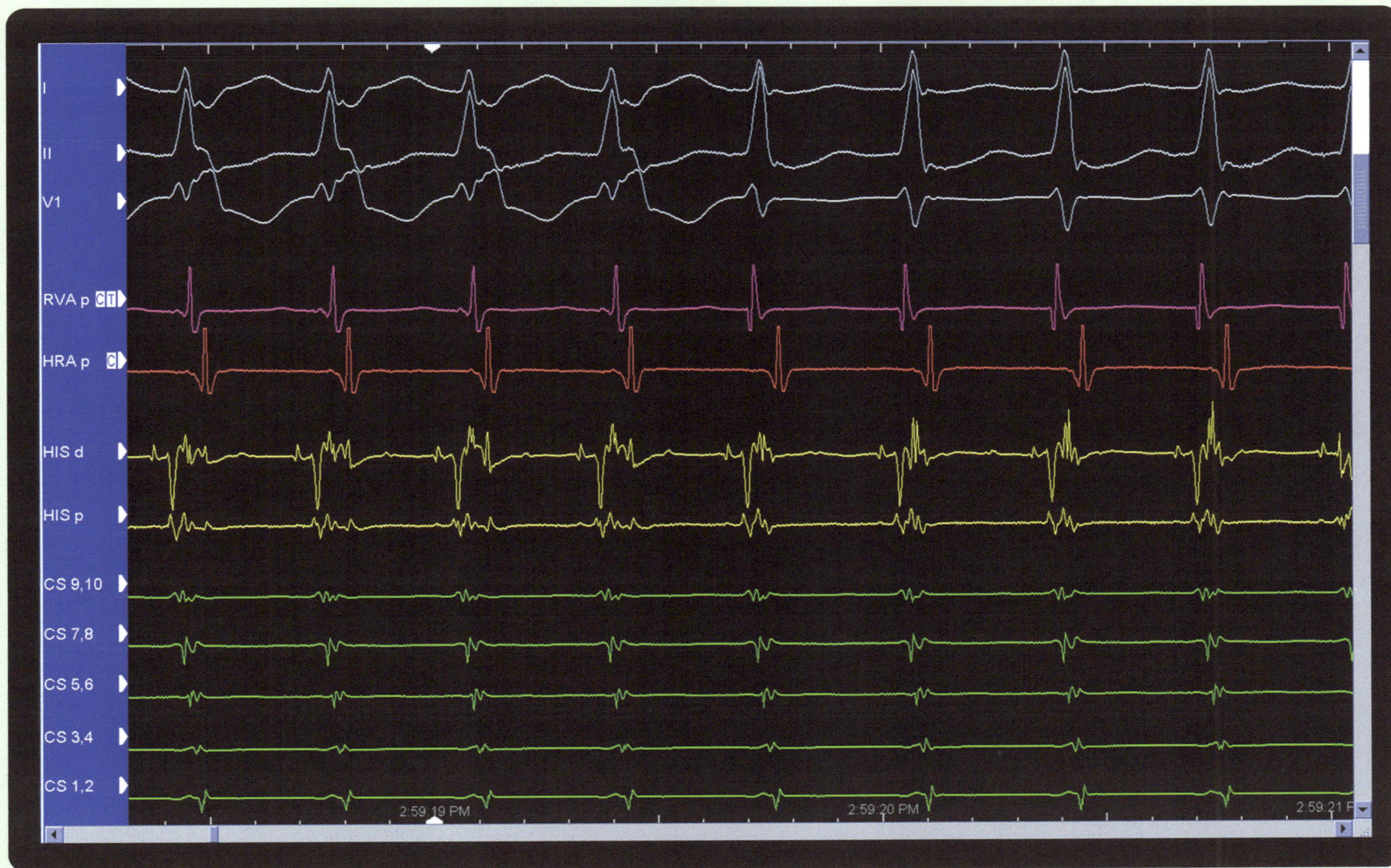

First Beat

1. Blocks in Rt Bundle
2. Continues down Lt Bundle
3. Trans-septal conduction
4. Retrograde up Rt Bundle
5. Collides in proximal Rt bundle

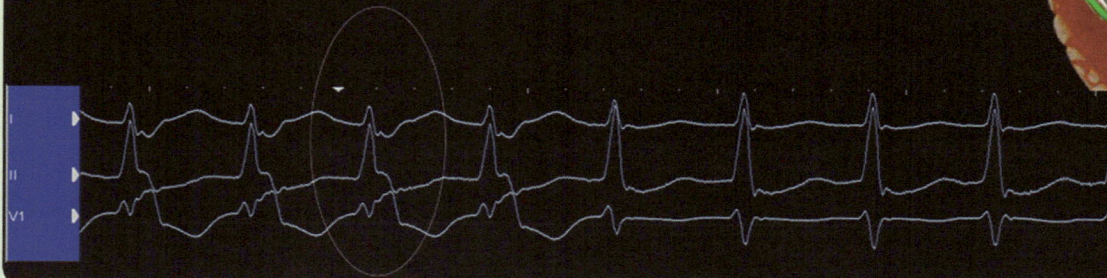

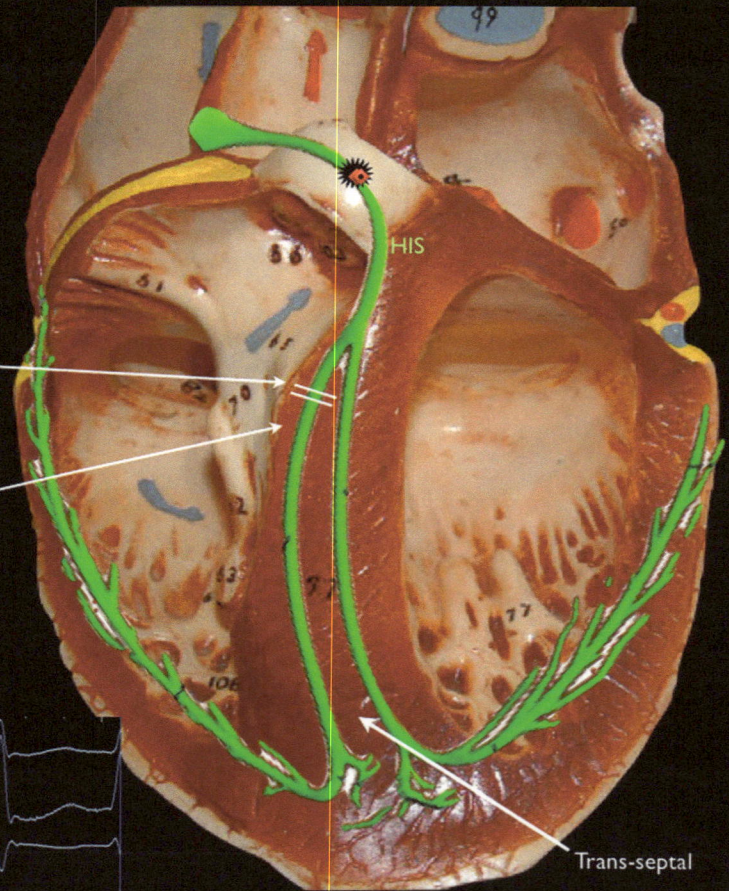

Trans-septal

HIS

Figure 1

Discussion

Spontaneous normalization of the QRS occurs due to the progressive shortening of the refractory periods of the bundle branches, as well as enhanced conduction velocity in response to an increase in heart rate and/or catecholamine levels.

As mentioned previously, this is a form of functional block and is a normal physiological phenomenon. In this example, a RBBB spontaneously resolves. Note the intermediate QRS morphology just before complete normalization.

In **Figure 1**, the initial RBBB occurs quite proximally in the bundle. The wave of depolarization continues to conduct down the left bundle, through the interventricular septum and retrograde up the right bundle. The collision point between the antegrade and retrograde waves is quite proximal in the right bundle.

continues

Second Beat

1. Rt Bundle shortens its ERP and enhances conduction
2. Trans-septal conduction continues
3. Collision point more distal in Rt bundle

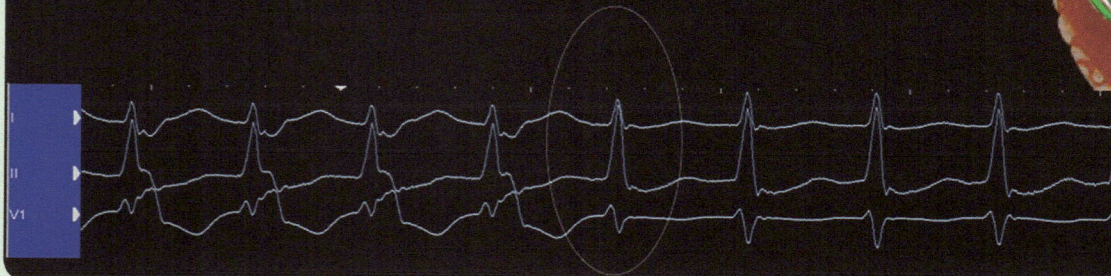

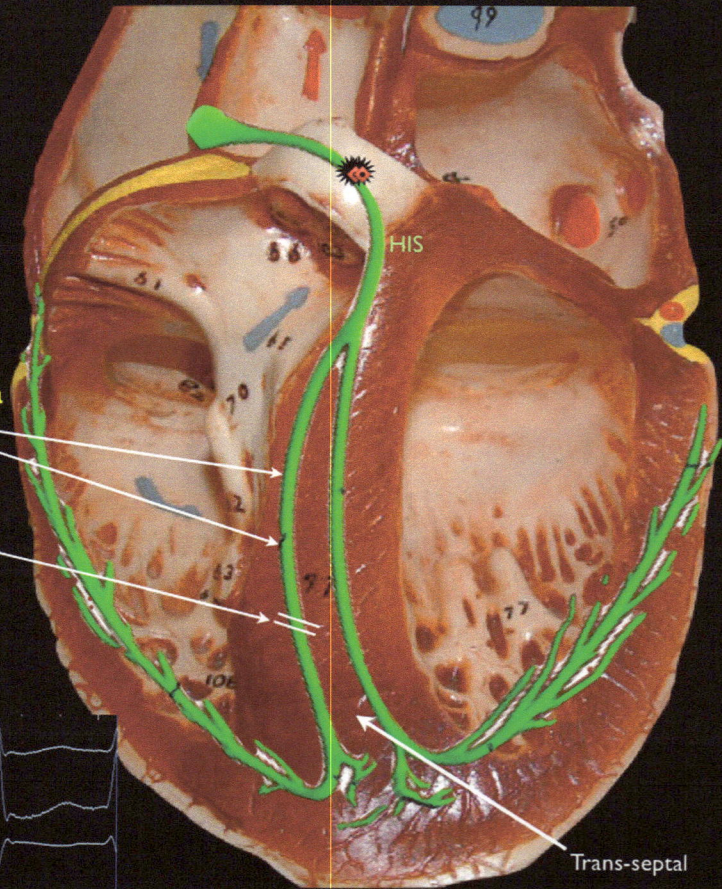

Trans-septal

HIS

Figure 2

In the next beat of tachycardia (shown in **Figure 2**), the collision point of the 2 waves is more distal in the right bundle. With each subsequent beat, the point of collision occurs progressively more distally in the right bundle.

continues

Third Beat

1. Rt Bundle has completely adapted
2. Normal conduction resumes

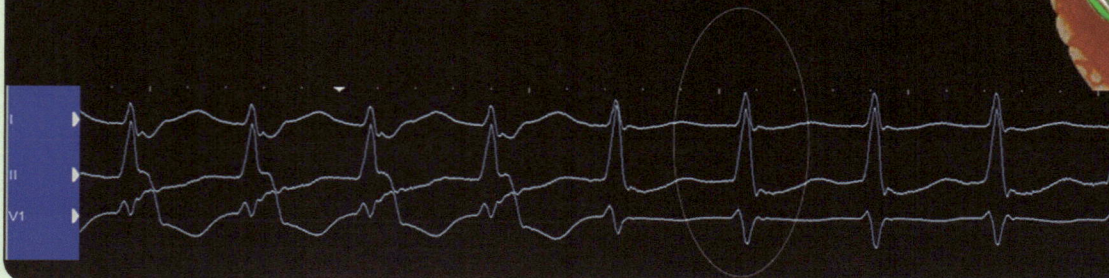

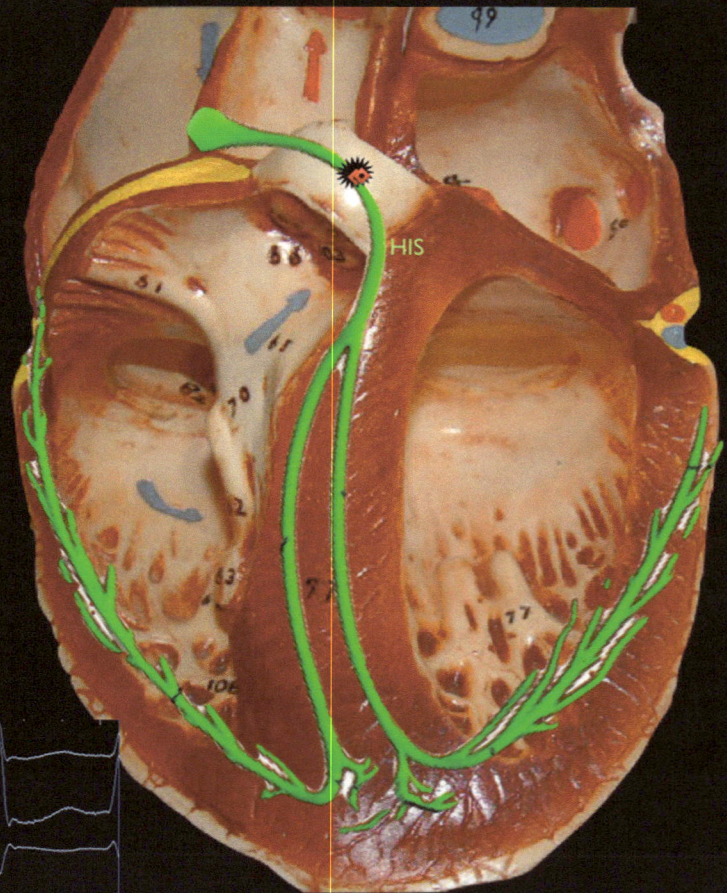

Figure 3

When the collision occurs *distal to* the insertion site of the right bundle, the QRS normalizes (**Figure 3**).

[Presented as movie: https://bit.ly/purvesadvanced.]

1.6 Bundle Reset

As in the case displayed previously in Section 4, we have ongoing AVNRT. Once again, the physiology of the bundle branch block (aberrancy) is the focus. If spontaneous normalization does not occur, can we "force" normalization?

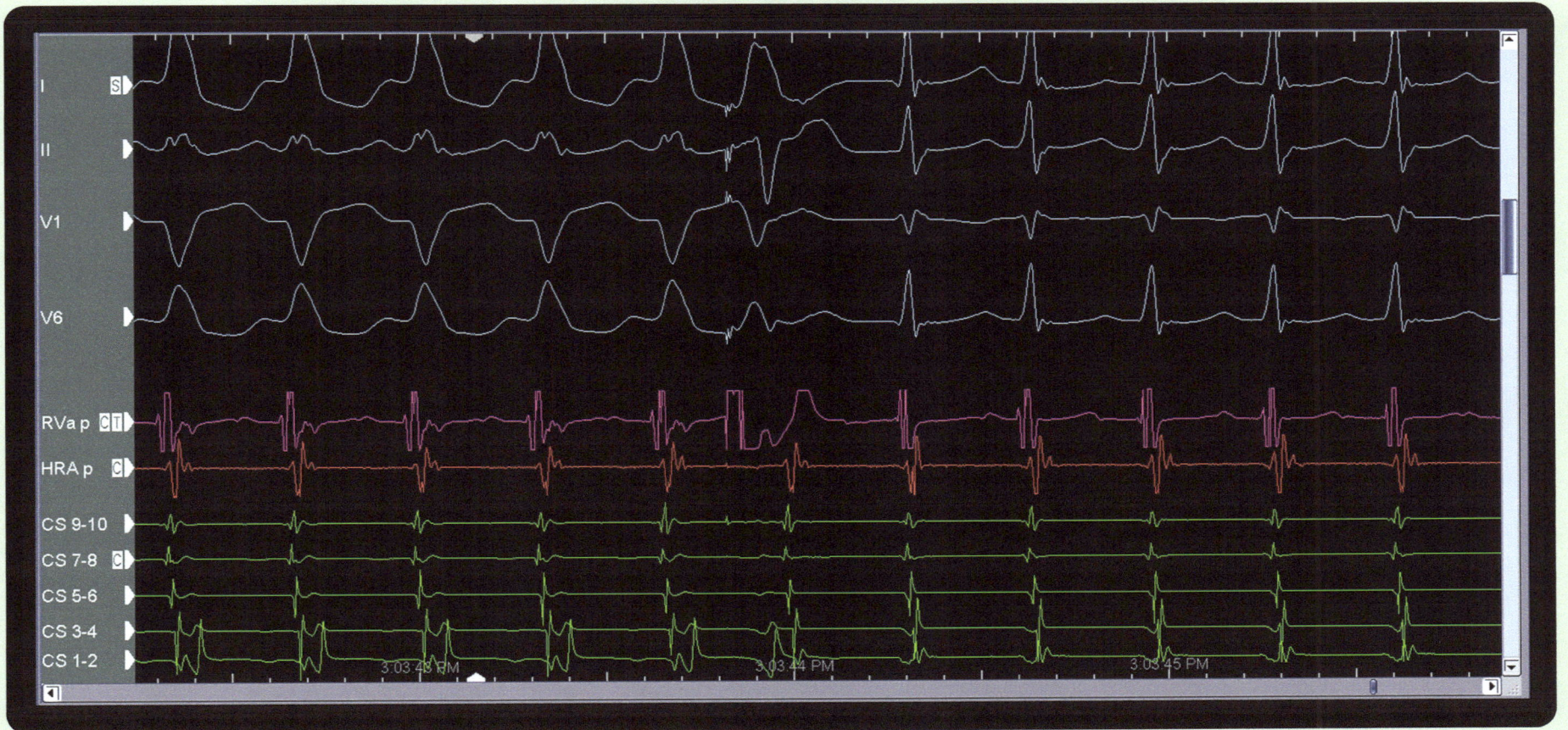

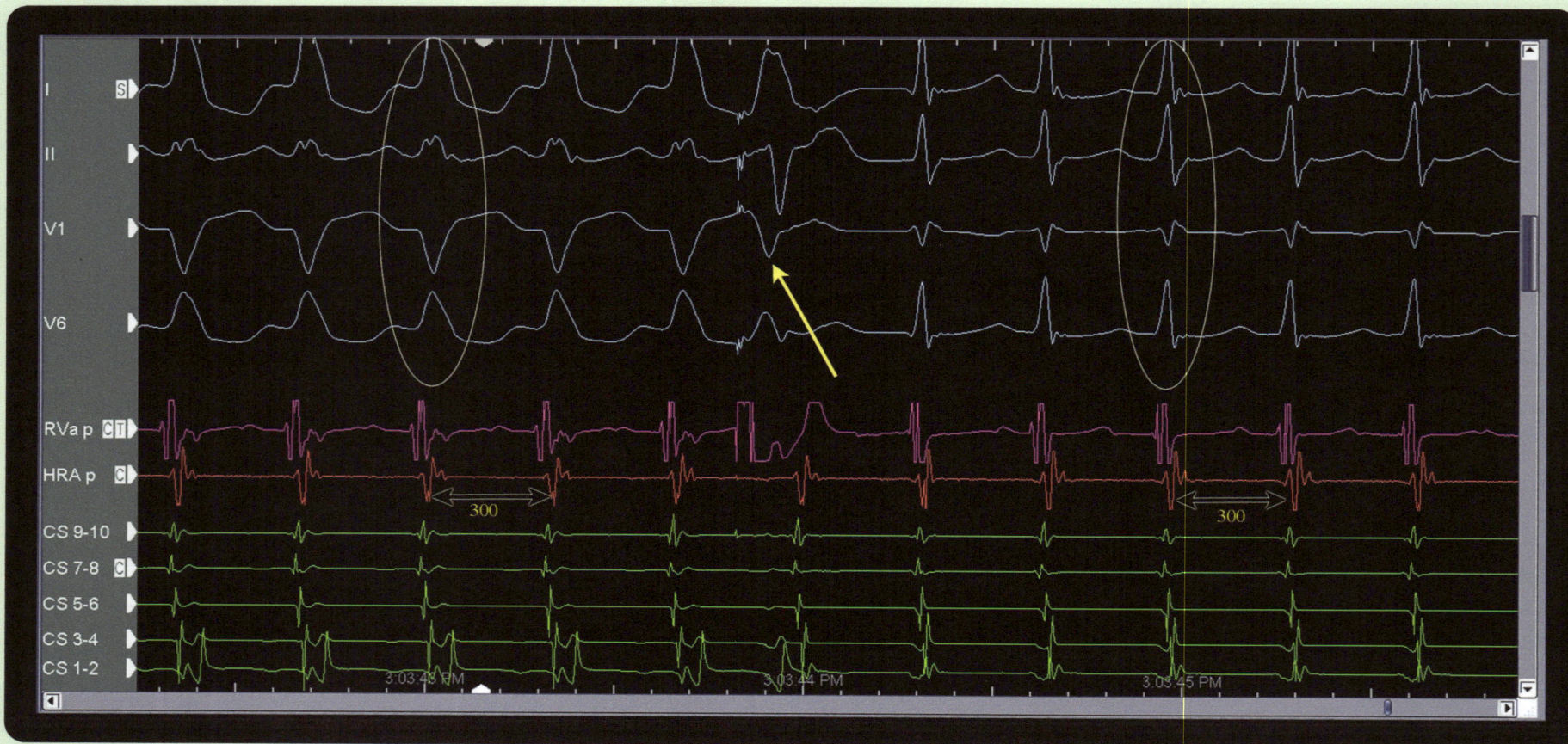

Discussion

In this tracing, we see a QRS transition from a LBBB pattern to a normal narrow QRS morphology following a ventricular extrasystole (or premature ventricular contraction (PVC)) from the RVA catheter.

The right ventricular PVC resolves the LBBB by preexciting both bundle branches. This shortens their refractory periods and also allows them more time to recover before the next beat of the AVNRT tachycardia arrives.

This is also known as "resetting" of the bundle branches.

1.7 Differentiation

This is a nice example of a very common observation. The question to be answered is, "How do you differentiate between 2 AV nodal echoes versus 2 junctional beats?"

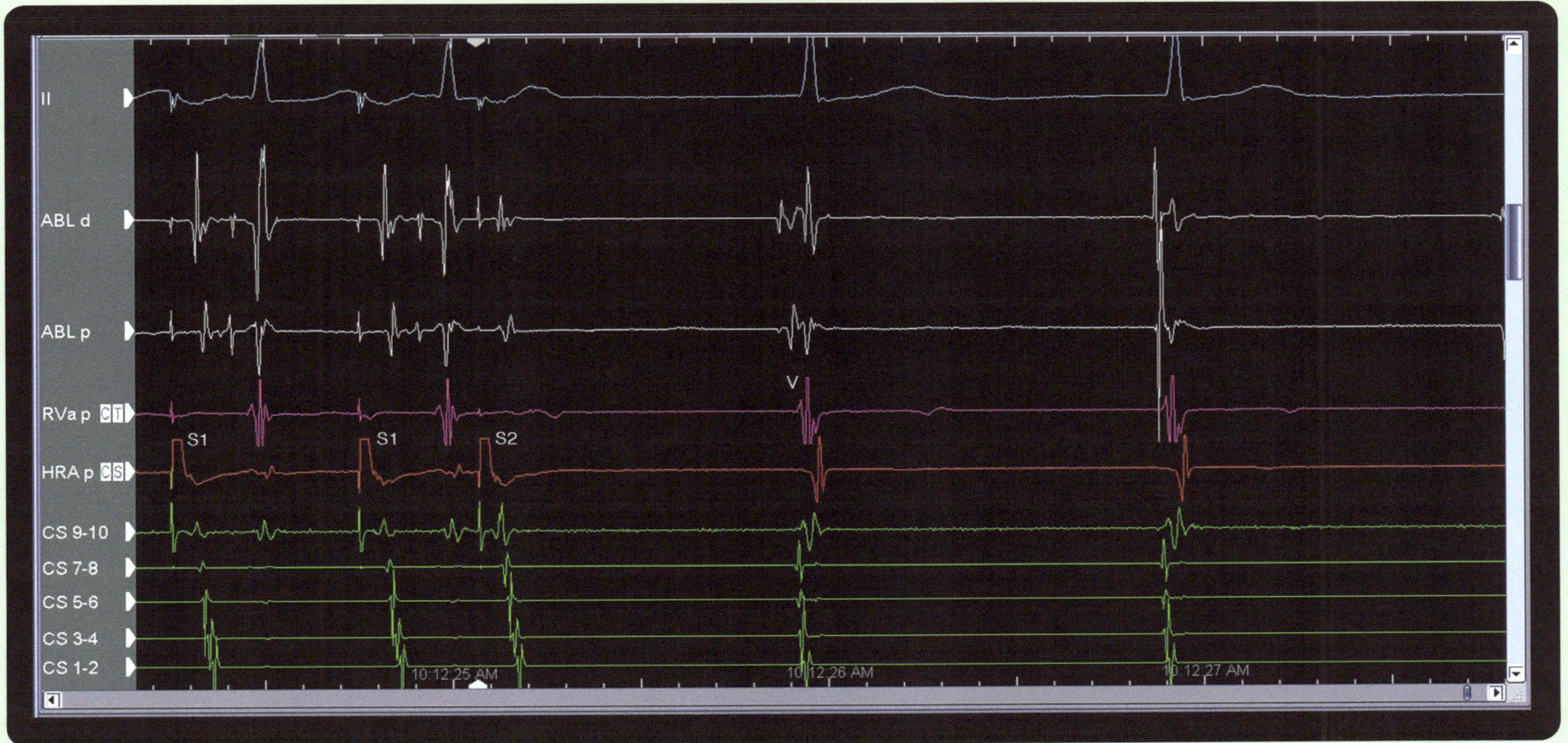

Figure 1

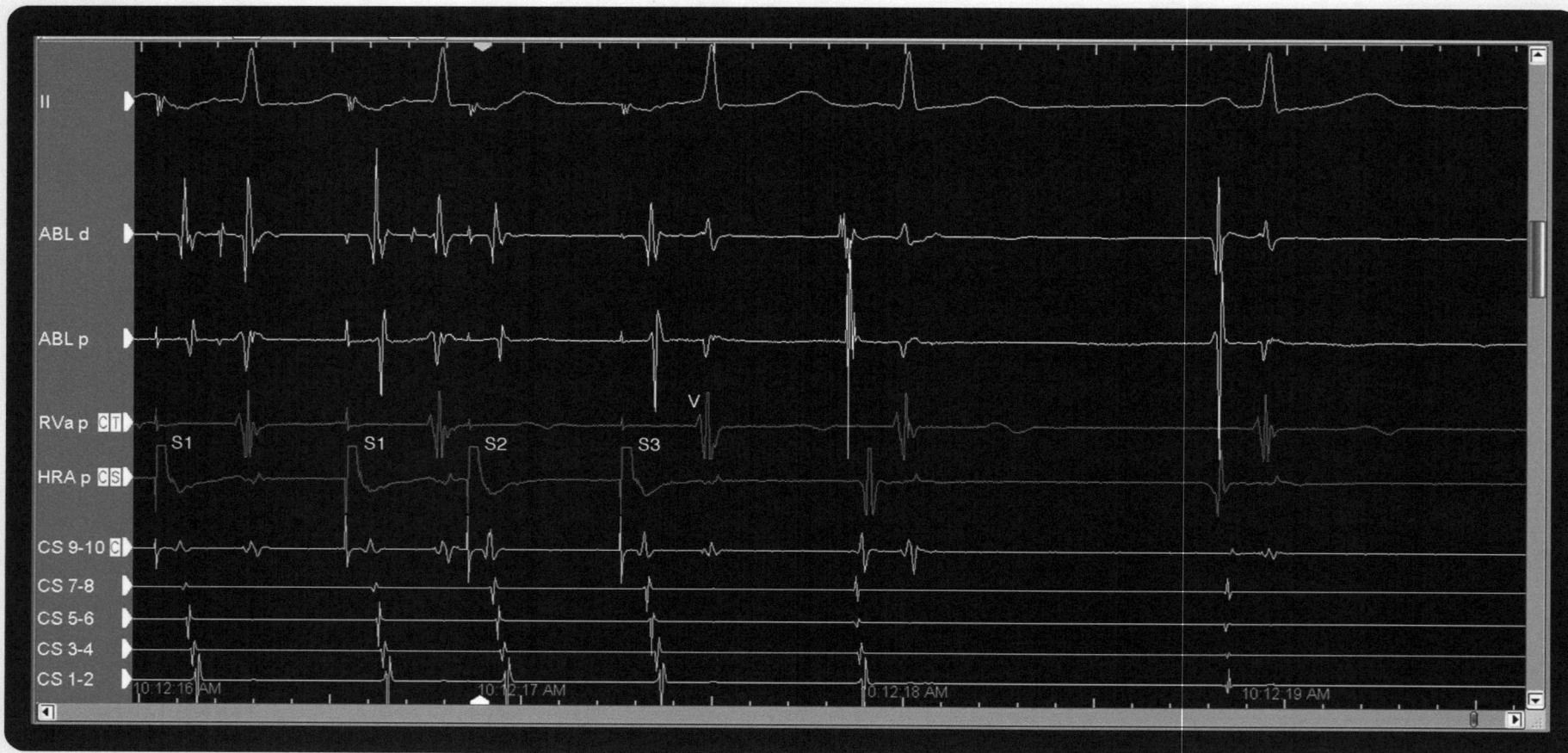

Figure 2

Discussion

Figure 1: Following an atrial extra-stimulus (S1S2), there is a long pause followed by 2 beats with simultaneous activation of the atrium and the ventricle. Atrial activation is concentric, and the V-A time is almost zero.

There are 2 possibilities for this observation: 1. The S2 is conducted down a slow pathway and retrograde up a fast pathway resulting in 2 sequential AV nodal echoes or 2. The S2 blocks and the 2 subsequent beats are junctional.

Figure 2: In this figure, an S3 is added after the same S1S2 we see in Figure 1. In this sequence, the pause following the S2 is interrupted by the S3, which appears to conduct.

There are also 2 possible explanations for this response:

1. The S2 is conducted down a slow pathway and the S3 blocks. If this were the case, the S3 would not be able to penetrate the AVNRT circuit since the antegrade fast pathway would be refractory and the slow pathway is being used. Therefore, the timing between the S2 and the next QRS should be the same as we see in Figure 1.

Alternatively...

2. The S2 blocks and the S3 conducts down the fast pathway. If this were the case, the S3 should advance the next QRS (i.e., it should occur earlier than it did in Figure 1).

continues

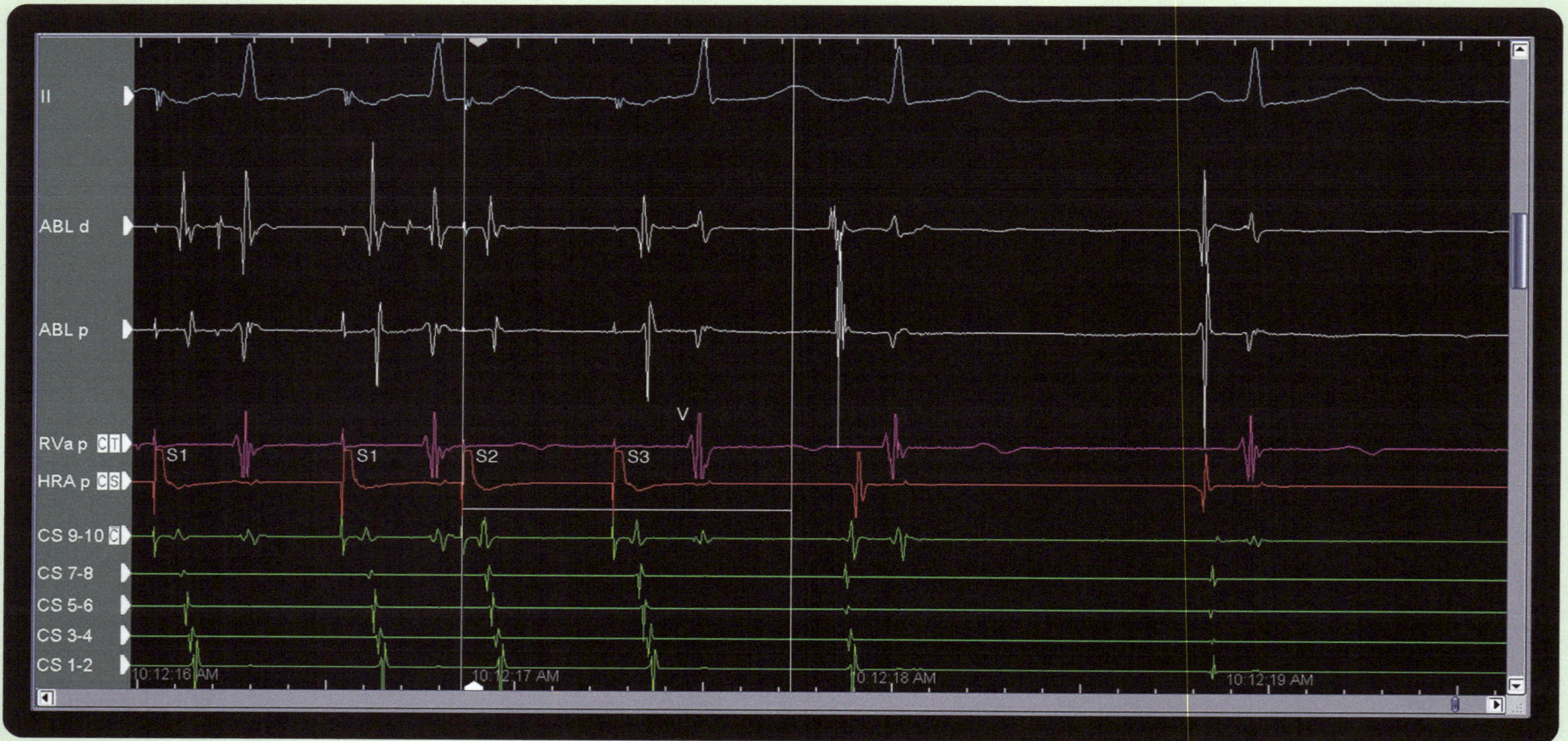

Figure 3

Figure 3 provides the answer. In this figure, calipers mark the interval between the S2 and the first QRS seen in Figure 1. As you can see, the QRS clearly occurs earlier than it did in Figure 1, suggesting that it was advanced by the S3.

This suggests that the S2 likely blocked, and that the last 2 beats in Figure 1 are actually junctional beats and not AV nodal echoes.

1.8 Concealment

This figure nicely demonstrates the concept of concealed conduction.

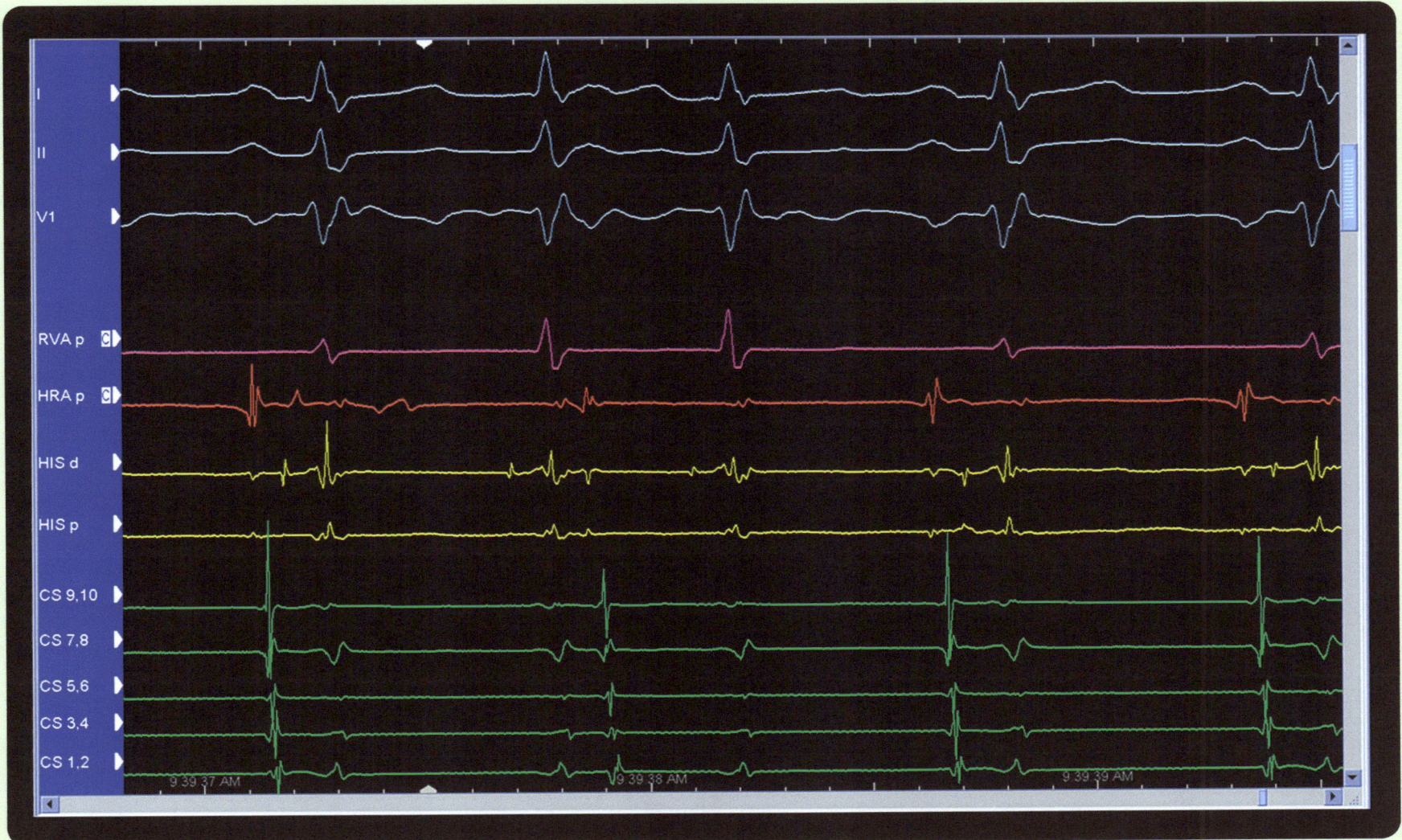

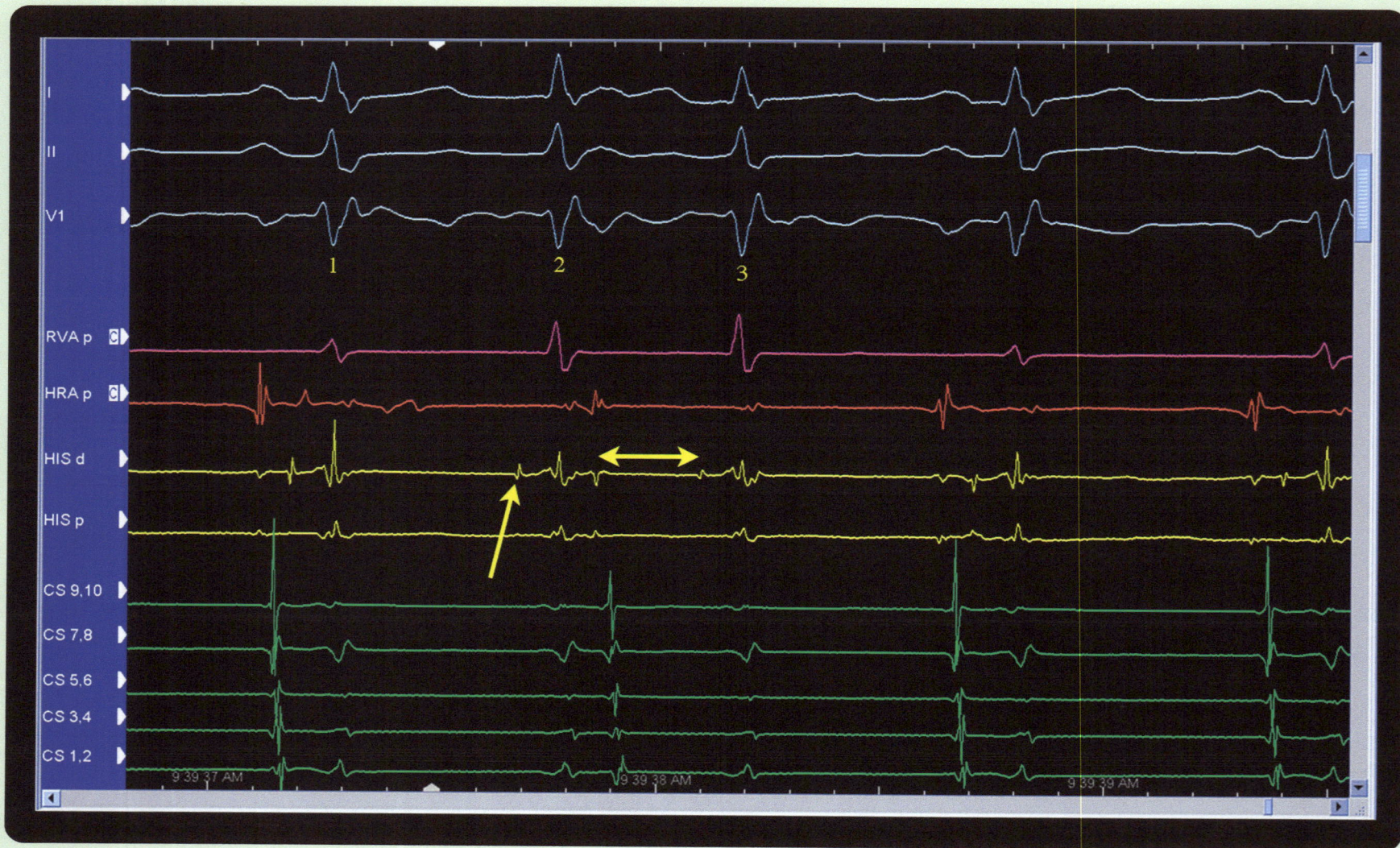

Discussion

The first beat in this tracing shows a normally conducted sinus beat, followed by a junctional beat (with the His deflection being first), and then a sinus beat conducted with a long A-H interval.

The prolonged A-H interval on the third beat is due to "concealed" retrograde conduction into the AV node from the preceding junctional beat. This means that the wave of depolarization generated by the junctional beat retrogradely penetrated the AV node (just as an extra-stimulus would), causing the AV node to decrement when the next sinus beat arrives. In other words, the junctional and sinus beats in sequence, produced the same response as a tightly coupled "S1S2" from either the atrium or the ventricle.

Another possible explanation is that the junctional beat could have concealed into the AV nodal fast pathway, rendering it refractory, thus forcing the sinus beat to conduct over a slow pathway. Dual AV nodal physiology would need to be demonstrated during the EP study to make this scenario possible.

continues

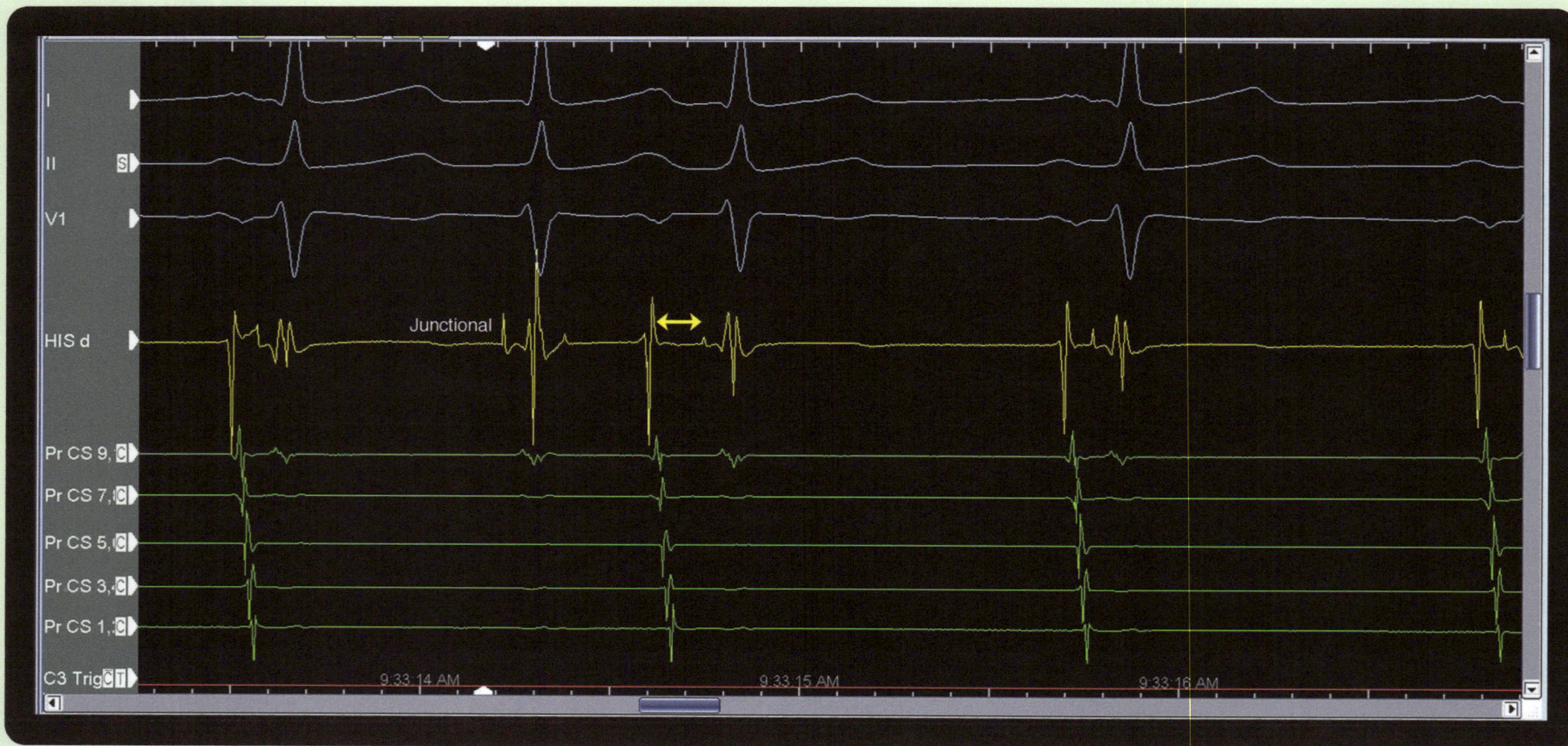

Figure 3

Figure 3 is another example demonstrating a similar AV nodal response, that is, a prolonged A-H following a junctional beat.

1.9 Concealment 2

This figure nicely demonstrates the concept of concealed conduction into an accessory pathway.

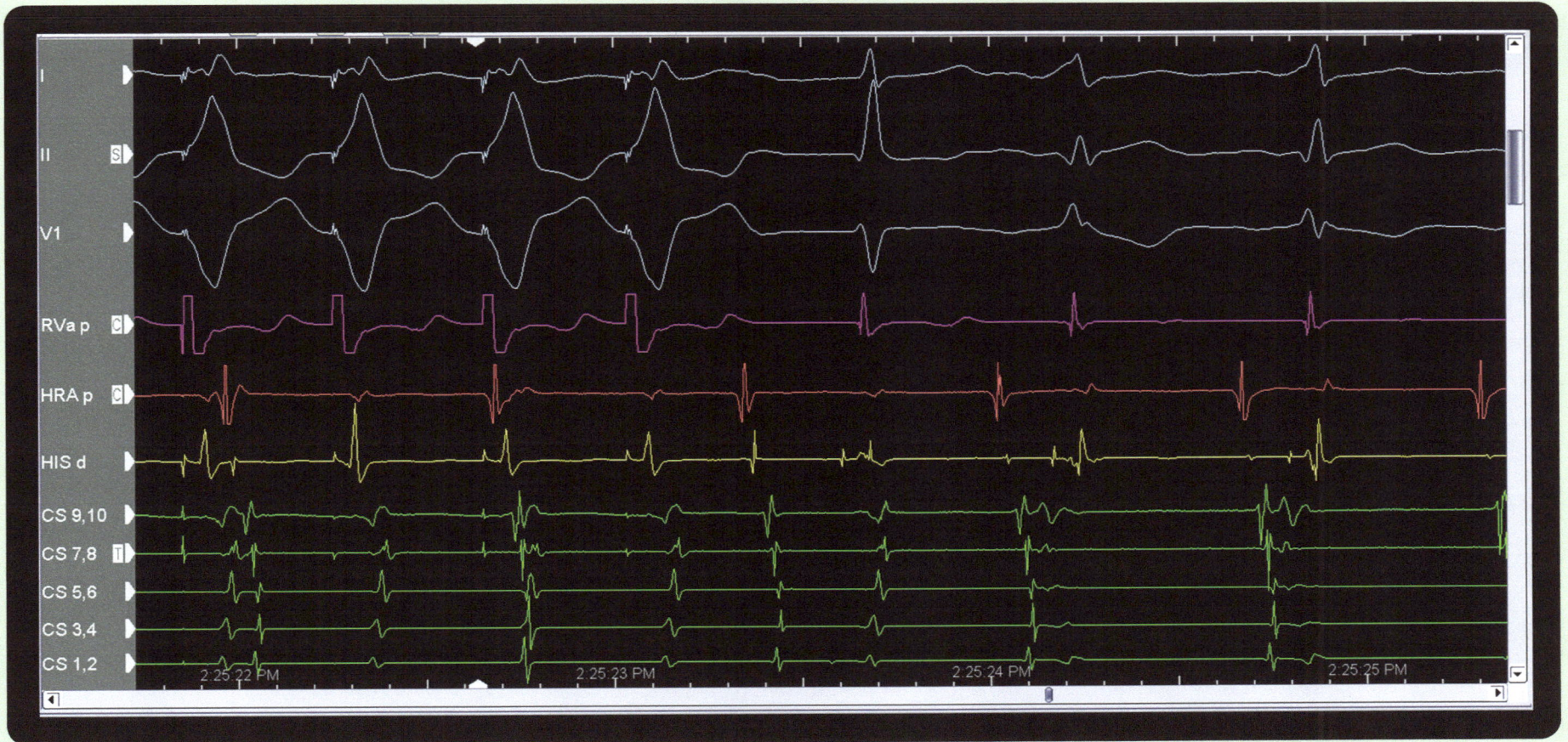

Figure 1

47

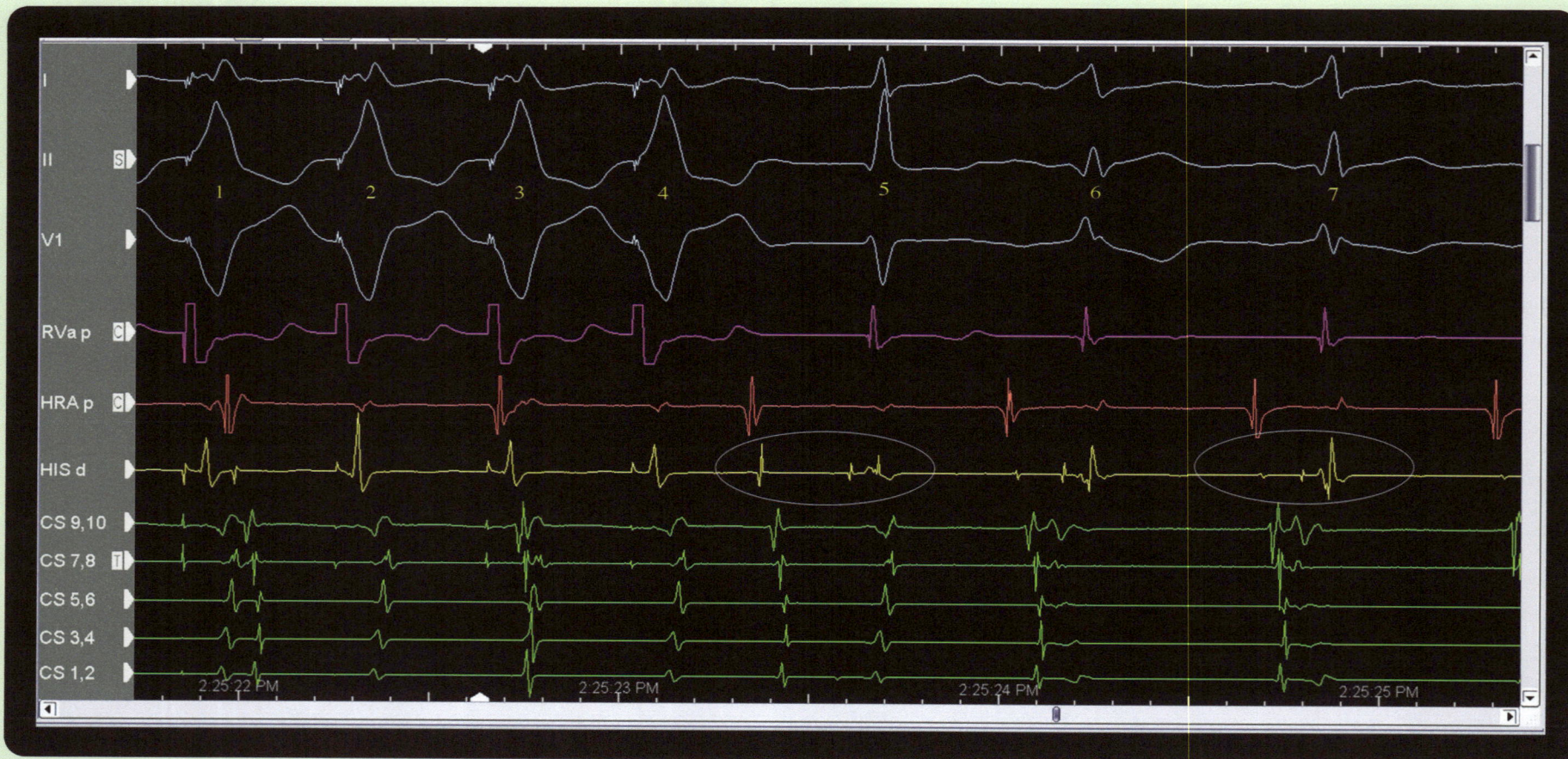

Figure 2

Discussion

Draw your attention to the last 2 beats on **Figure 1** (#6 and #7). Based on the short A-V interval and presence of a delta wave on the surface ECG, the patient is preexcited.

Now examine the first 4 beats.

Pacing from the RVA catheter demonstrates no V-A conduction, suggesting that there is block in the AV node and in the AP at this pacing rate.

After RVA pacing ceases, a sinus beat conducts with a long A-H interval and a narrow QRS (beat #5). Why is this beat not preexcited?

Although there is no V-A conduction, the AV node and AP still receive retrograde input from the RV pacing. This "concealed conduction" into the AV node causes it to decrement, making the subsequent sinus beat conduct with a longer A-H interval. Simultaneously, concealed conduction into the accessory pathway renders the AP refractory to the oncoming sinus beat, which is why the QRS #5 is narrow.

Interestingly, beats #6 and #7 demonstrate slightly different QRS morphologies due to variable degrees of preexcitation.

Concealed conduction into the AV node or the AP can change their conduction properties without electrogram or electrocardiographic manifestations.

1.10 Escape Rhythm

This example demonstrates an atrial S2 that blocks in the
AV node. Following is a sinus beat and a *very* short A-V interval.
How do you explain beat #3?

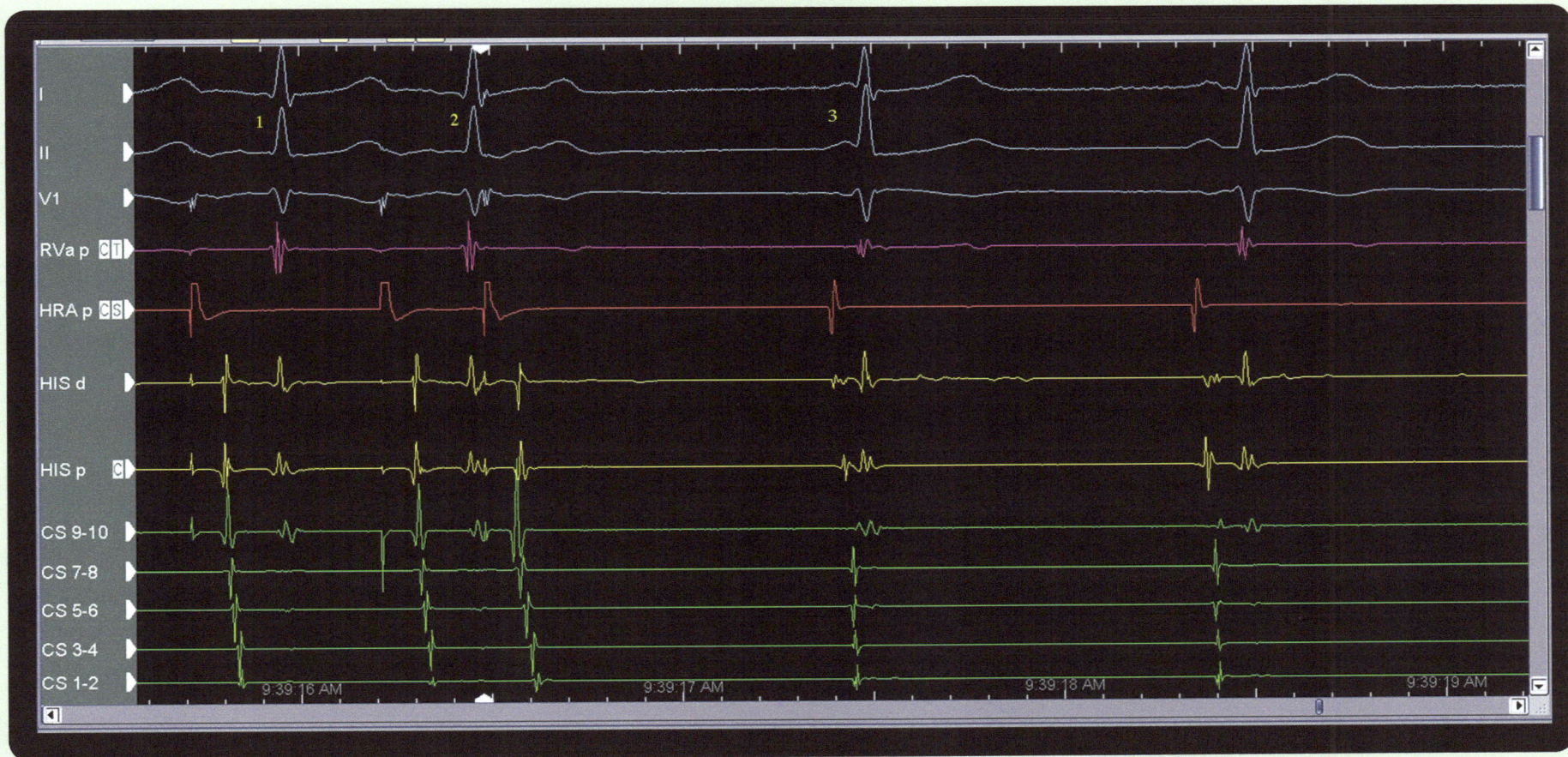

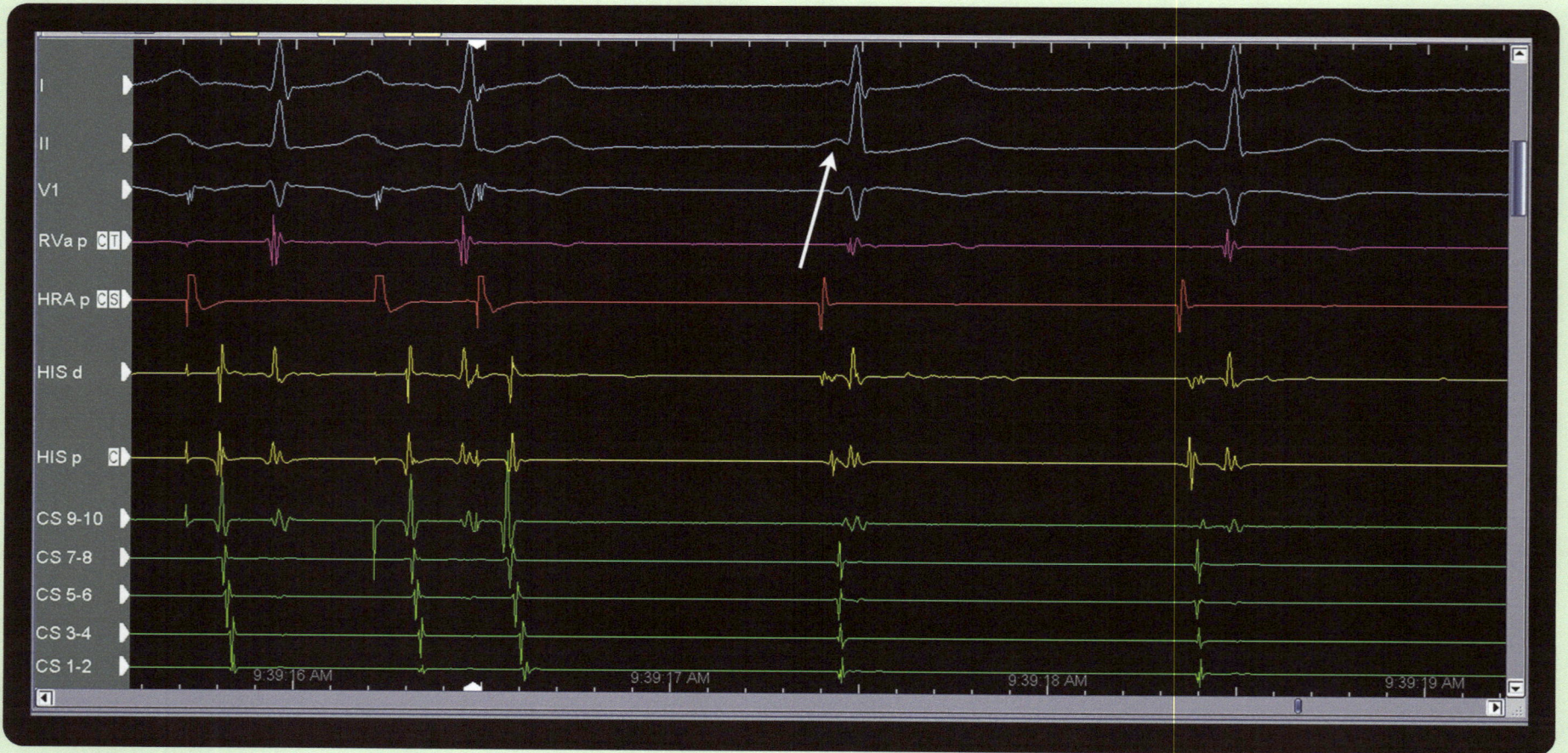

Discussion

The PR interval associated with this first sinus beat is physiologically too short to conduct down the AV node. Therefore, beat #3 must be a junctional beat that occurred at the same time as the sinus beat. The oncoming sinus wave of depolarization could not conduct through the AV node since the junction and ventricles have already been depolarized and are therefore refractory.

This is an example of a junctional escape beat. This occurred due to overdrive suppression of the sinus node from programmed atrial extra-stimulus pacing. Recovery of the sinoatrial (SA) node from this overdrive suppression was long enough to allow the junctional escape rhythm to take over.

1.11 Nodal Function

This tracing demonstrates atrial extra-stimulus pacing with preexcitation on the surface ECG. What is the AV node doing?

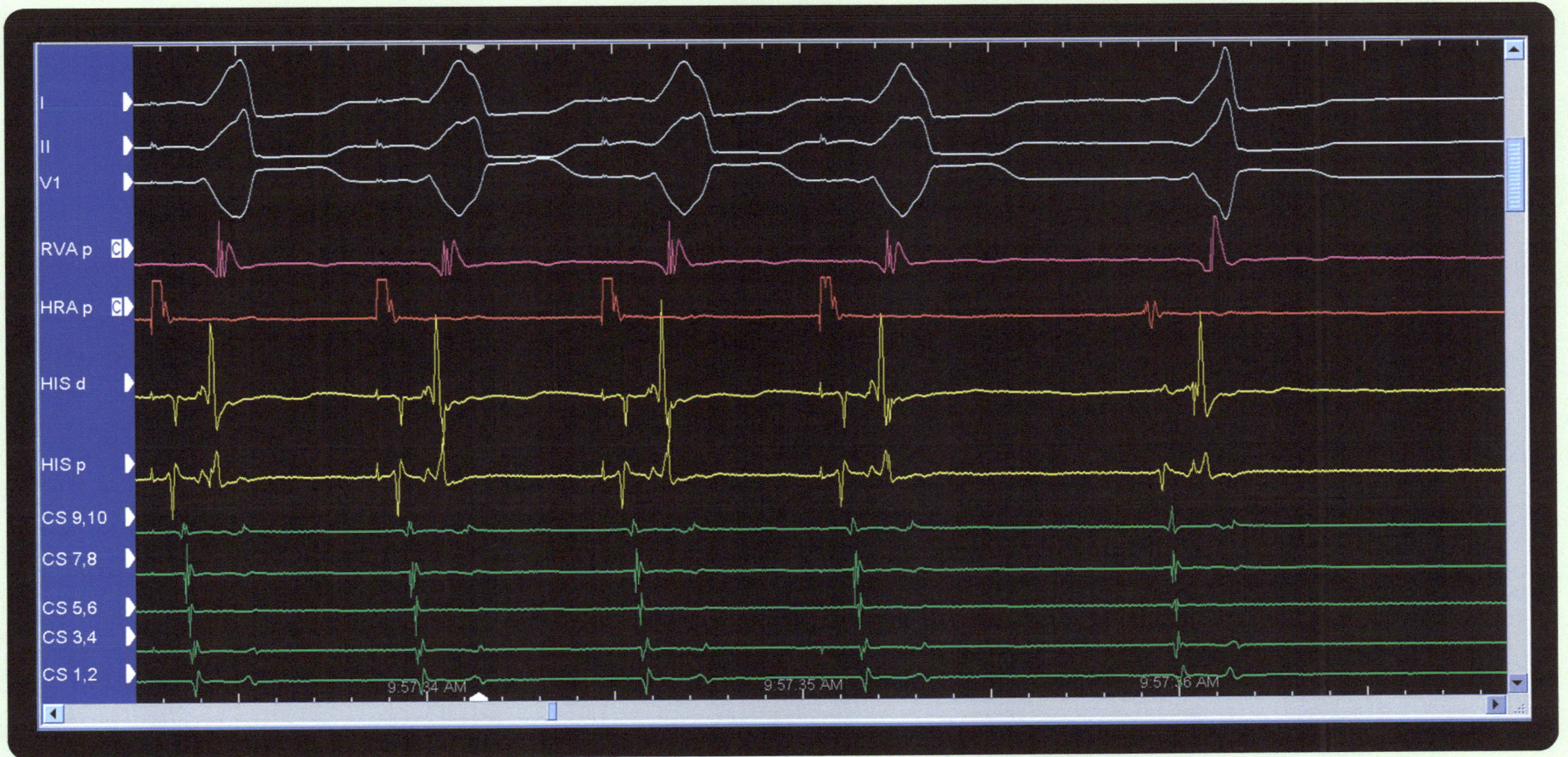

Figure 1

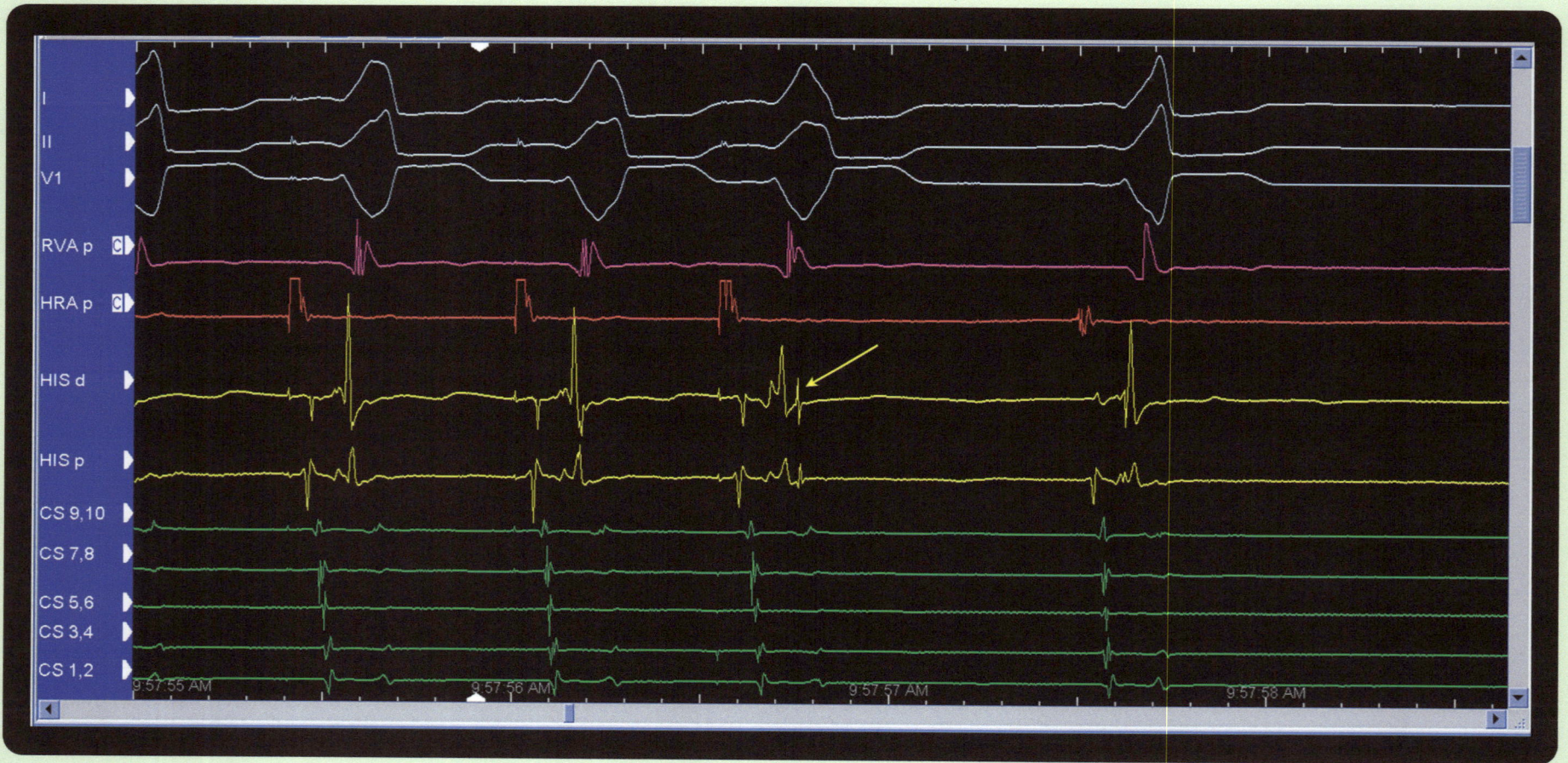

Figure 2

Discussion

Notice also that the intracardiac electrograms show that the His "V" electrogram occurs before the local ventricular electrogram recorded at the RVA catheter. This suggests that the ventricle is being activated at the base before the RV apex (where the His-Purkinje and right bundles activate the ventricle). This is intracardiac evidence of preexcitation.

In **Figure 2**, the arrow points to a His deflection that occurs *after* the local His "V" electrogram. Why does this happen?

Recall that, unless the patient is in AV reentrant tachycardia (AVRT) (either orthodromic or antidromic), there is always fusion of conduction between the AV node and the AP. Therefore, as the atrial extra-stimulus pacing is delivered earlier and earlier, the AV node will progressively decrement, but the AP usually will not. Therefore, the A-V interval will remain constant, but the A-H interval will continue to prolong. The local His bundle electrogram may eventually be seen *after* the local His "V" electrogram.

Notes

Cardiac Electrophysiology 2
CHAPTER 2   AVNRT

2.1 Signal ID

The first step in analyzing any tracing involves identifying the atrial (A), ventricular (V), and His bundle (H) electrograms (EGM) where appropriate. In some tachycardias, this identification may be difficult.

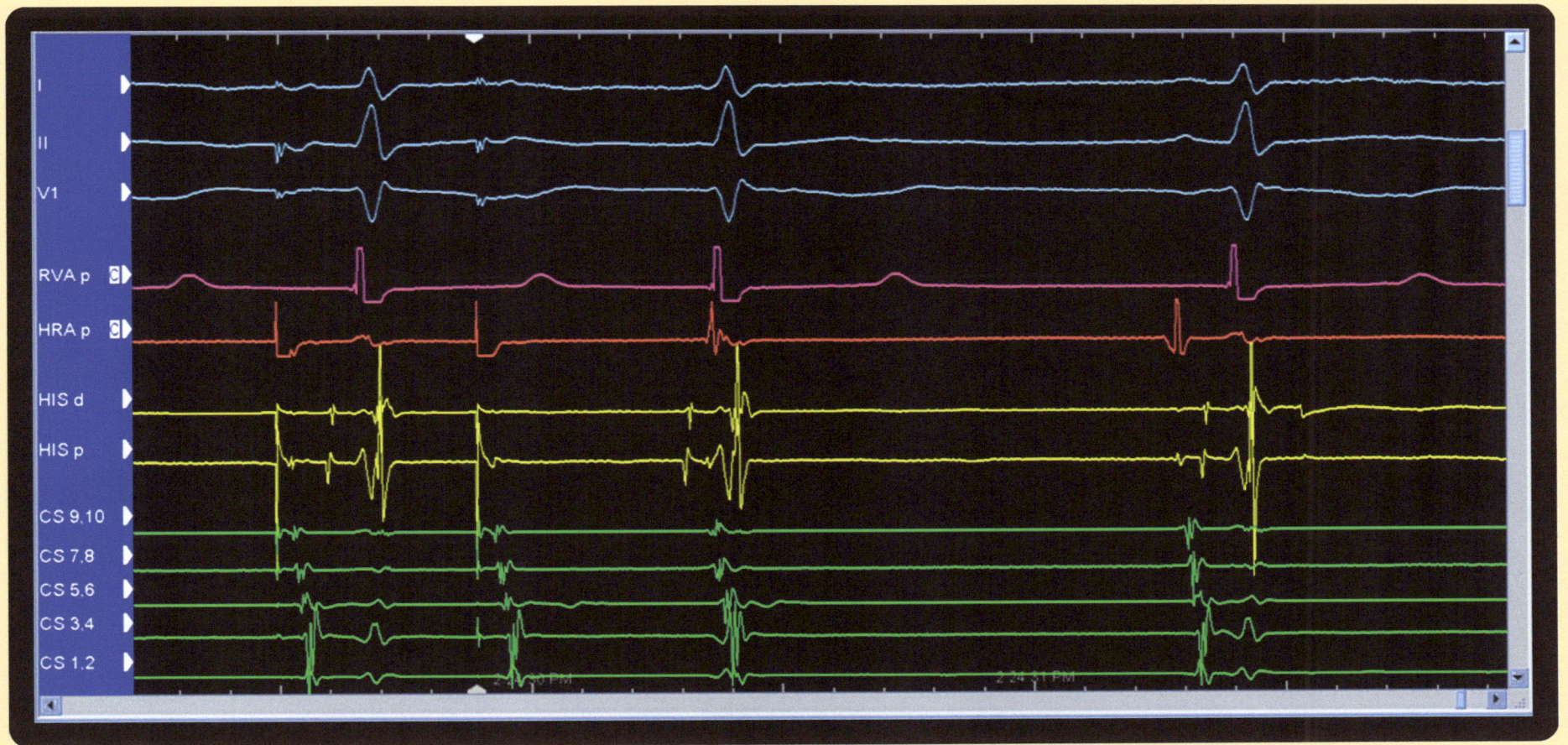

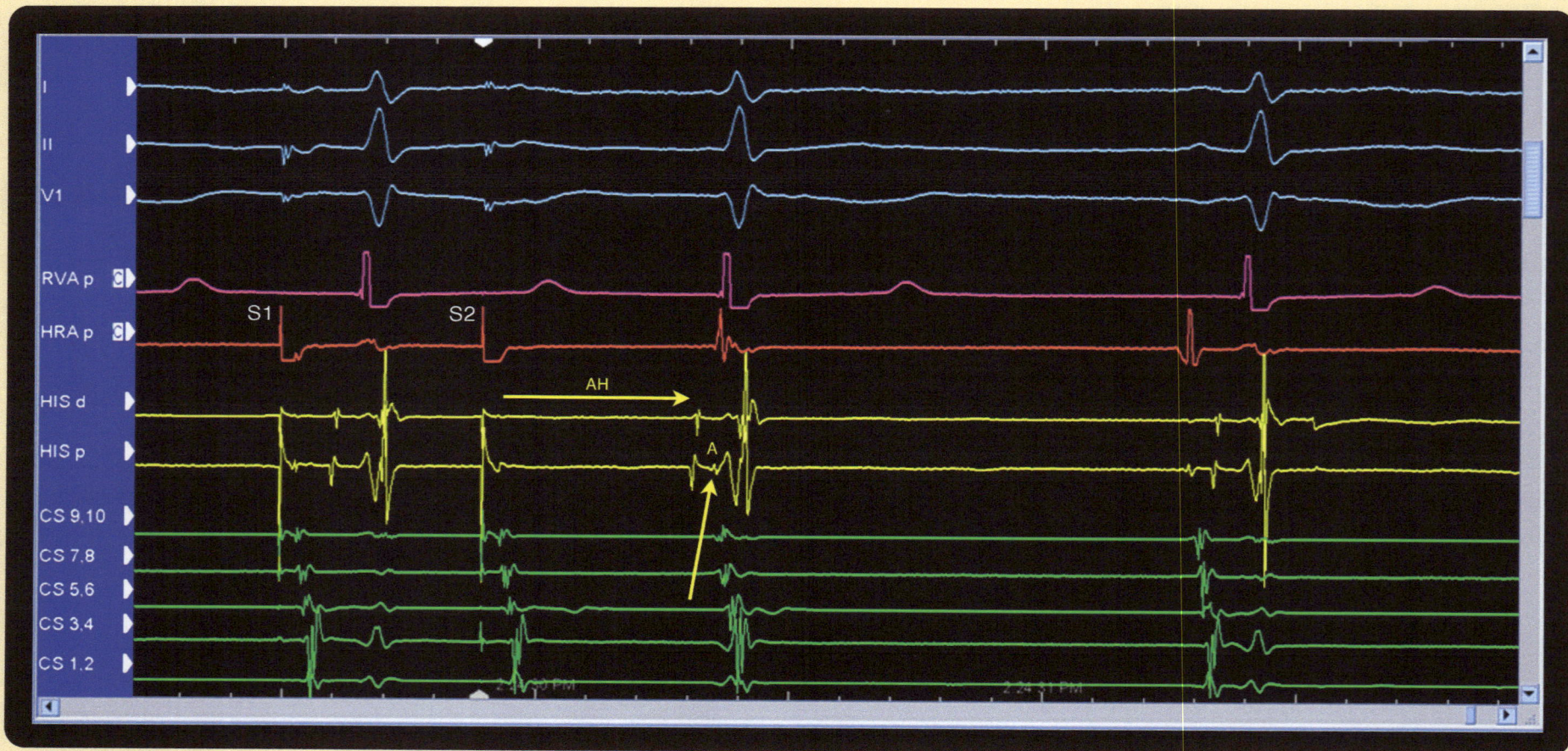

Discussion

This tracing shows atrial extra-stimulus testing with the last of the S1 drive train and the S2 displayed. The S2 generated an AV nodal echo beat, as evidenced by the presence of the unstimulated "A" in the high right atrium (HRA) channel.

In typical AVNRT, the S2 propagates antegradely down the slow AV nodal pathway, through the His and on to the ventricles. Simultaneously, retrograde conduction back up the AV nodal fast pathway is occurring. The distance and conduction velocity of these competing activation wavefronts can sometimes result in the "A" preceding the "V," making correct identification of the "A" and the "V" electrograms more difficult.

When in doubt, place a caliper on the earliest onset of ventricular activation on any surface ECG lead. Most of the time (there are always exceptions), electrograms to the right of the caliper line are ventricular and electrograms to the left of the line are atrial. This is why measured V-A times in AVNRT are often zero or negative values!

2.2 Pathways

During any electrophysiology study for AVNRT, we identify the number of antegrade and retrograde AV nodal pathways by observing how many distinct populations of A-H and V-A intervals occur. How many pathways do you see in this tracing?

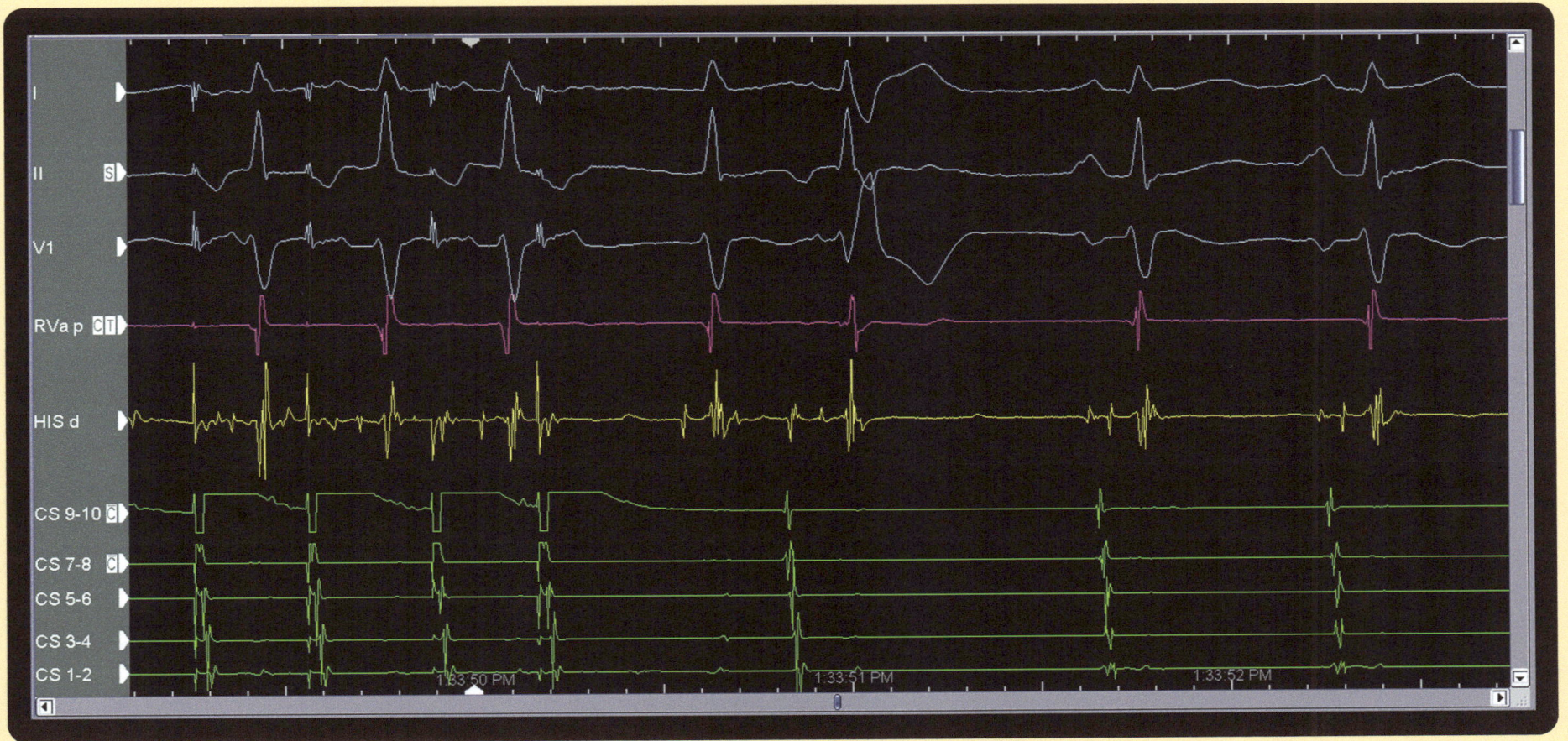

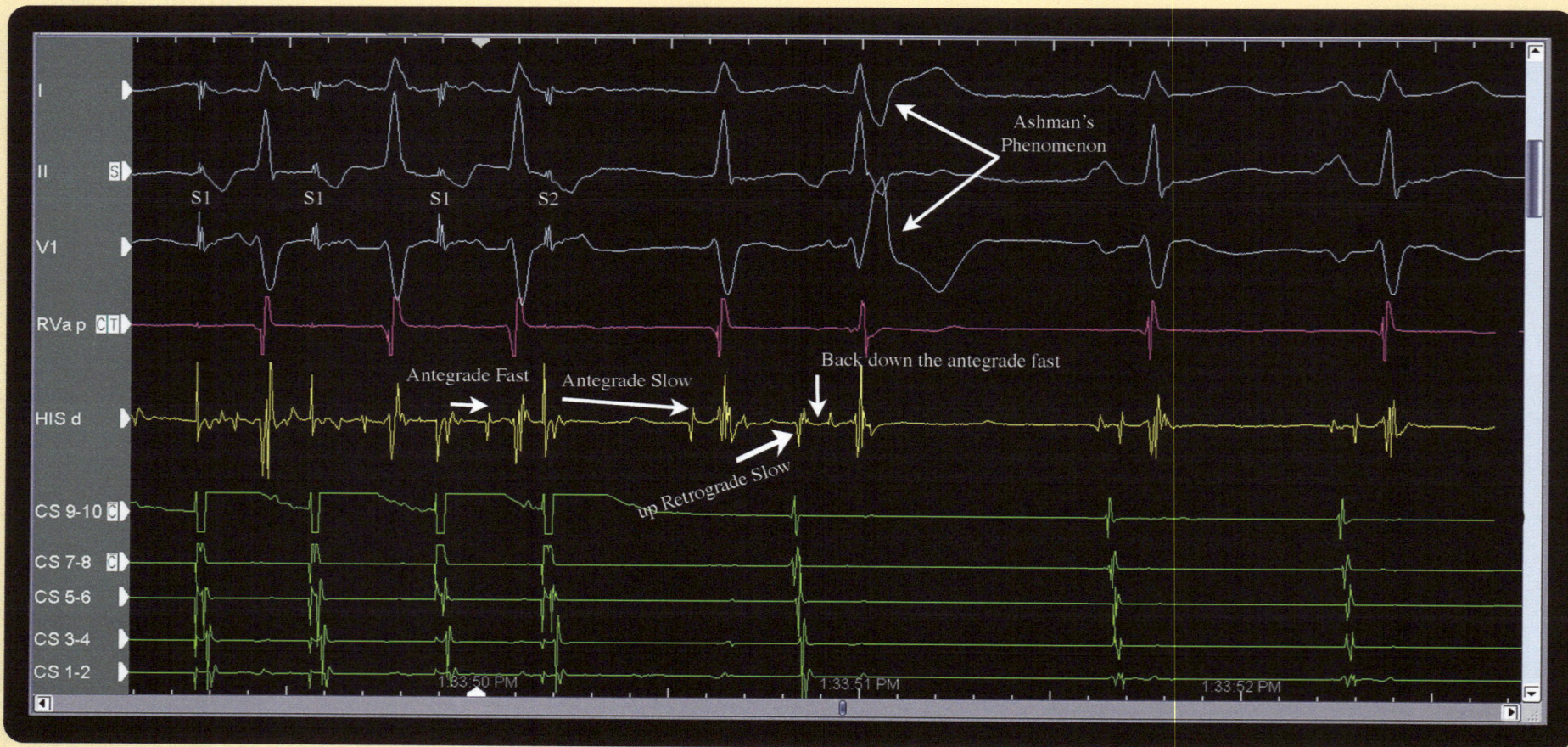

Discussion

In this tracing, extra-stimulus testing elicited 3 out of the 4 AV nodal pathways; 2 in the antegrade direction and 1 in the retrograde direction.

1. The S1's conduct to the His via an antegrade fast pathway with a short A-H interval.

2. The S2 conducts to the His via an antegrade slow pathway as evidenced by the much longer A-H interval.

3. Following the fourth QRS, retrograde conduction up an AV nodal slow pathway results in a retrograde "A" with a long V-A interval.

Not seen on this tracing, the patient also demonstrated a retrograde fast pathway.

Also of note in this tracing, the fifth QRS is the result of the AV nodal echo (back down the fast AV nodal pathway with a short A-H interval). The QRS morphology has a RBBB pattern due to Ashman's phenomenon.

2.3 Initiation

During a diagnostic study, utilizing S1S2 extra-stimulus pacing may result in some surprising tachycardias. This is a *very* unexpected result of ventricular extra-stimulus testing!

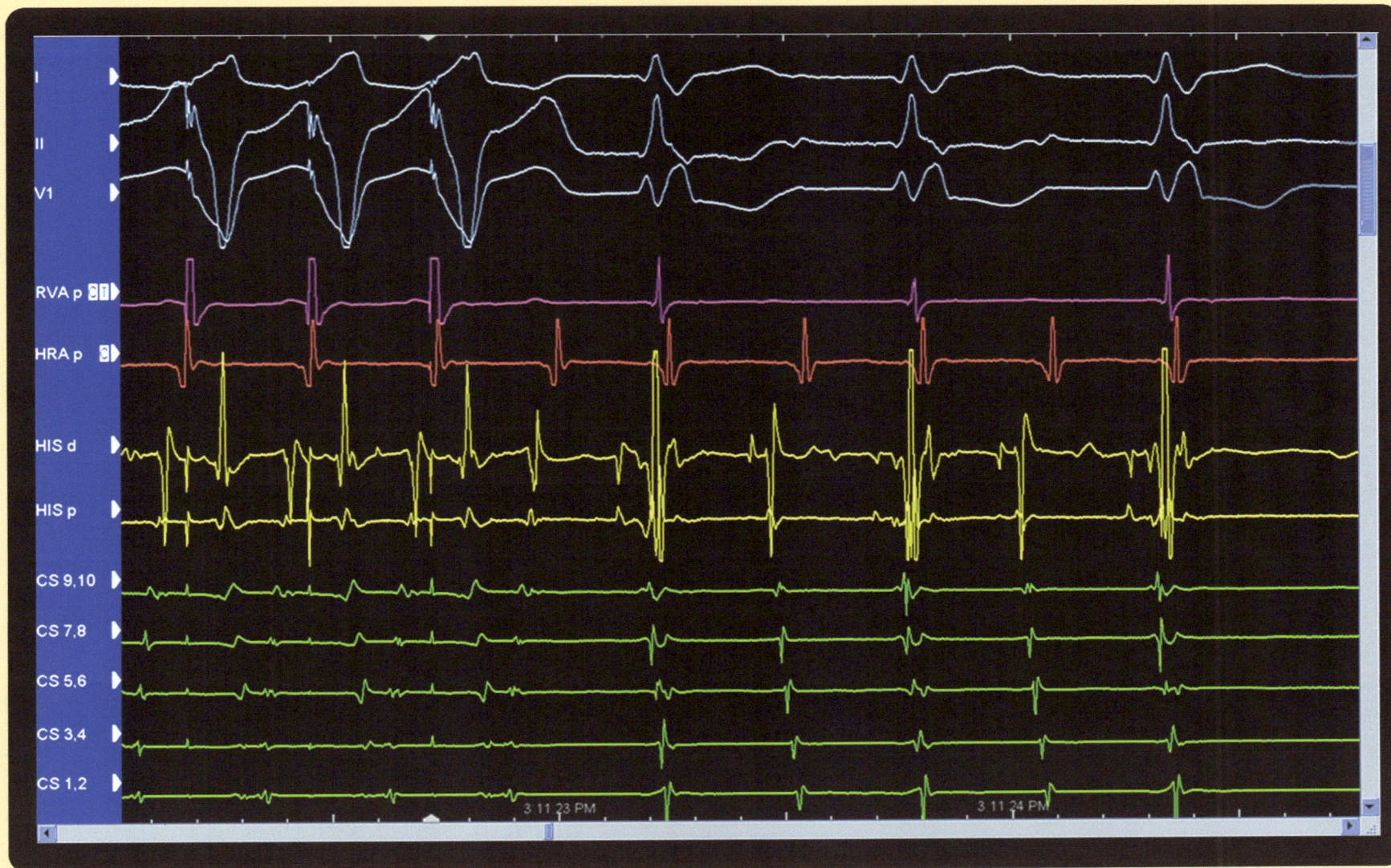

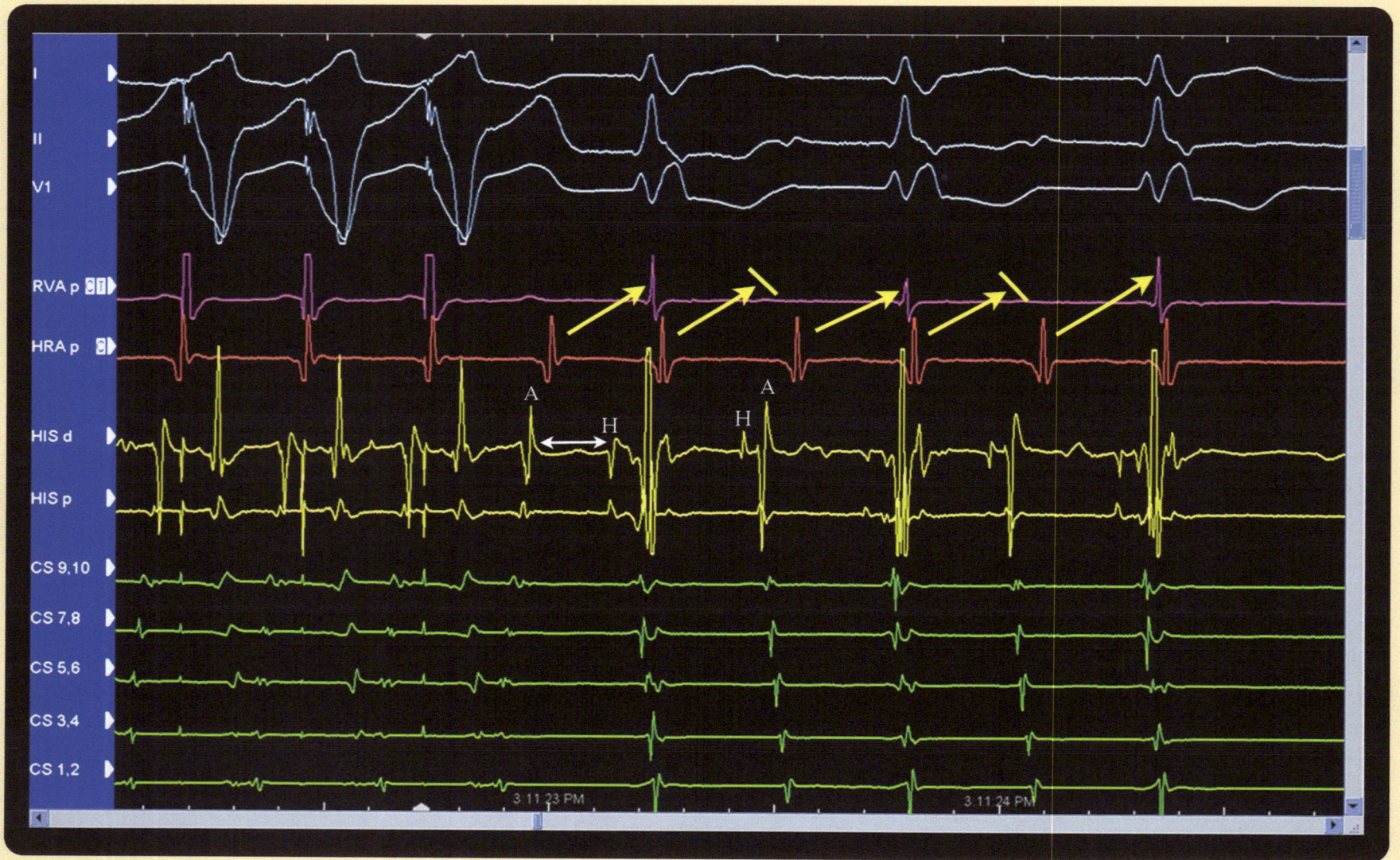

Discussion

With the S2, ventricular activation propagates to the atria via a left-sided AP (evidenced by the earliest "A" in CS 1-2). One would normally expect to initiate AVRT.

What ensues on the right half of this tracing, is a tachycardia where there are 2 "A"s for every "V." This immediately rules out AVRT, since AVRT requires the ventricles to maintain the tachycardia. Also, the V-A time on the beats where "V" and "A" coincide, is too short for AVRT.

Focusing attention on the His channel: Following retrograde atrial activation via the accessory pathway, the wavefront travels down an antegrade slow AV nodal pathway to the ventricle while simultaneously going back up a retrograde fast AV nodal pathway to yield another "A" (with a very short V-A time). We have initiated AVRT!

After initiation, the depolarization wavefront reenters the AV nodal slow pathway to the His bundle and back up the retrograde fast AV nodal pathway…but there is infrahisian block. You see a His deflection and another "A" but no "V" or QRS complex. On the ensuing AVNRT cycle and every second cycle thereafter, a ventricular response and QRS occurs.

We have initiated 2:1 AVNRT. This illustrates nicely that AVNRT can occur without any ventricular participation or activation.

2.4 Reset

Similar to the previous tracing, we have ongoing 2:1 AVNRT. Is it possible to convert 2:1 AVNRT into 1:1 AVNRT? Can you explain the mechanism by which this may be accomplished?

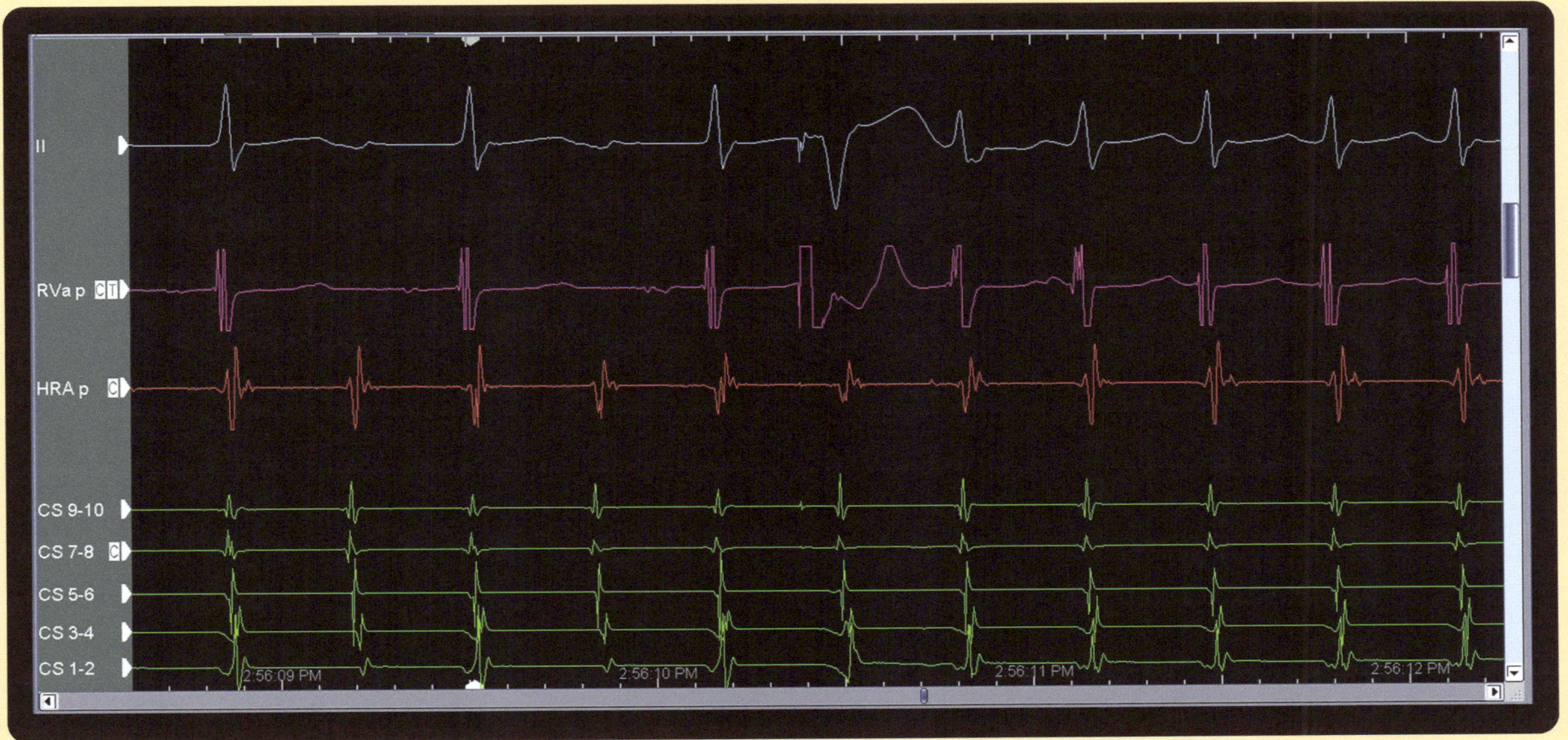

II

RVa p

HRA p

CS 9-10

CS 7-8

CS 5-6

CS 3-4

CS 1-2

2:56:09 PM   2:56:10 PM   2:56:11 PM   2:56:12 PM

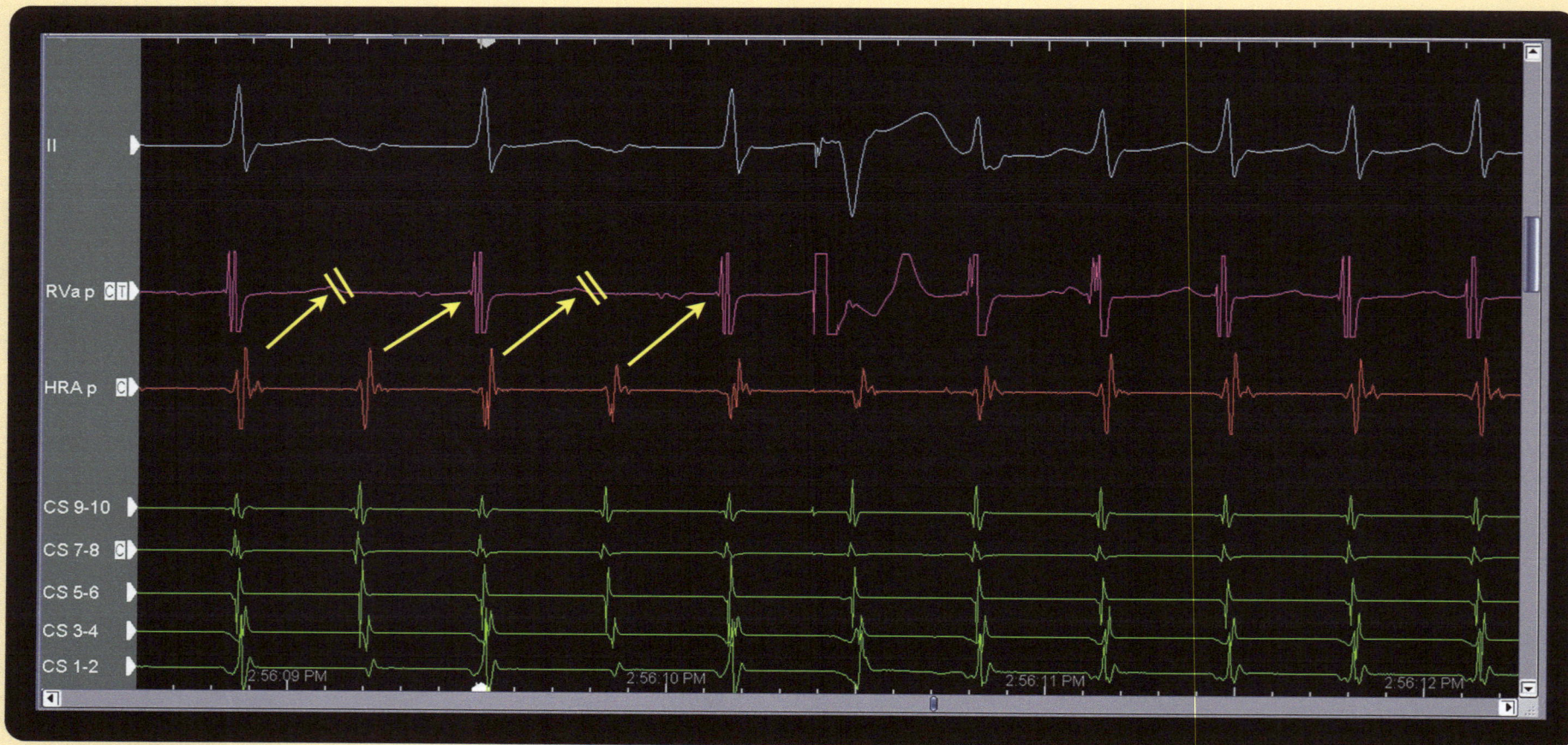

Discussion

The insertion of the paced ventricular beat (equivalent to a PVC) advanced or "reset" the refractory period of the His-Purkinje tissue below the AVNRT reentrant circuit. This results in the next and every subsequent AVNRT beat finding the His no longer refractory. Consequently, this converts the previous 2:1 A-V relationship into a 1:1 A-V relationship.

In response to an increase in heart rate, gradual shortening of the refractory period of the His-Purkinje tissue and enhanced conduction velocity, will often result in spontaneous resolution of the 2:1 A-V block. This is sometimes (somewhat colloquially) referred to as "peel back."

See Chapter 3, Section 12 for another example of spontaneous normalization.

2.5 Cool Initiation

In this tracing, we have just completed an S1S2 ventricular extra-stimulus drive. Describe what you see. Explain the origin of the last 3 beats.

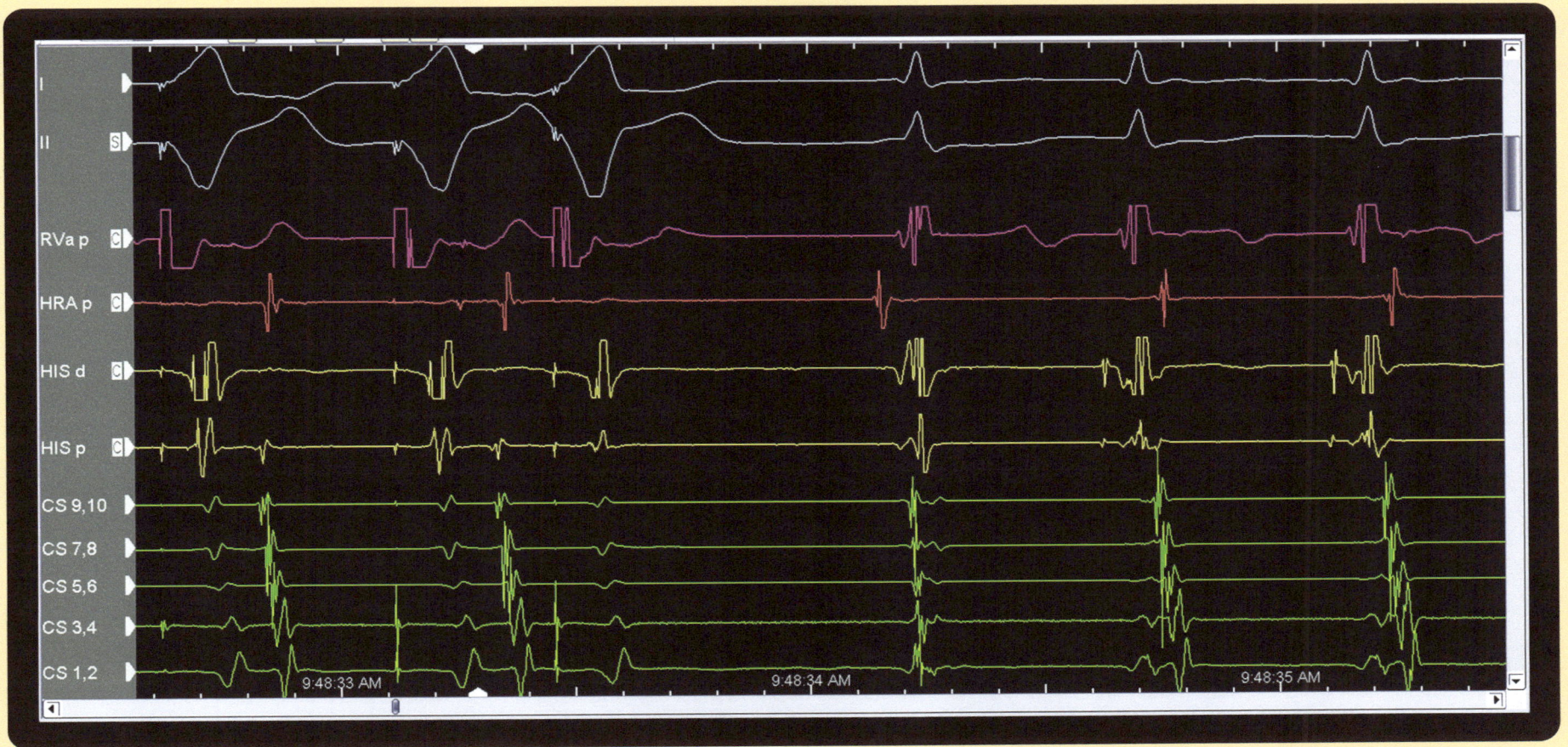

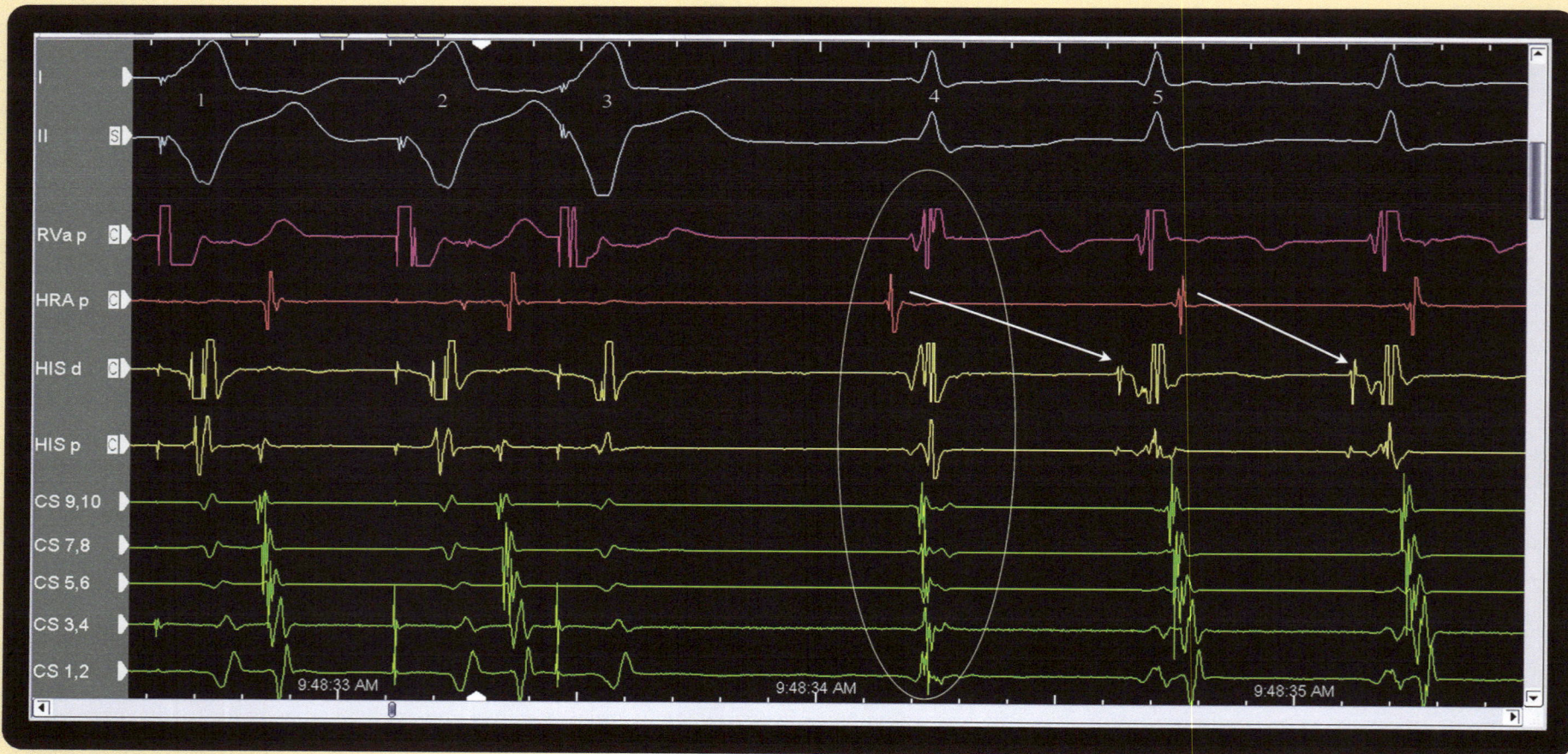

Discussion

QRS #4 is the interesting beat. A sinus beat has occurred (the "A" is earliest at the HRA). However, examining the surface ECG we see that the PR interval is too short to be physiologically possible. Therefore, QRS #4 must have been a coincidental spontaneous junctional beat. The key question here is: "How did the junctional beat contribute to the initiation of AVNRT?" (The last 2 beats in this tracing.)

The likely explanation is that the junctional beat must have conducted retrogradely (concealed) into the AV node rendering the antegrade fast AV nodal pathway refractory. The oncoming sinus beat would find only the antegrade slow pathway available (non-refractory). Therefore, conduction proceeds down the slow pathway, then subsequently back up the retrograde AV nodal fast pathway (with a short V-A interval) and initiates AVNRT (from beat #5 onward).

This is another example of the effect of concealed retrograde conduction into the AV node.

2.6 Block to A

As we have seen in previous tracings, AVNRT does not require the ventricles or the atria to sustain its reentrant circuit. What is the unusual observation in this tracing?

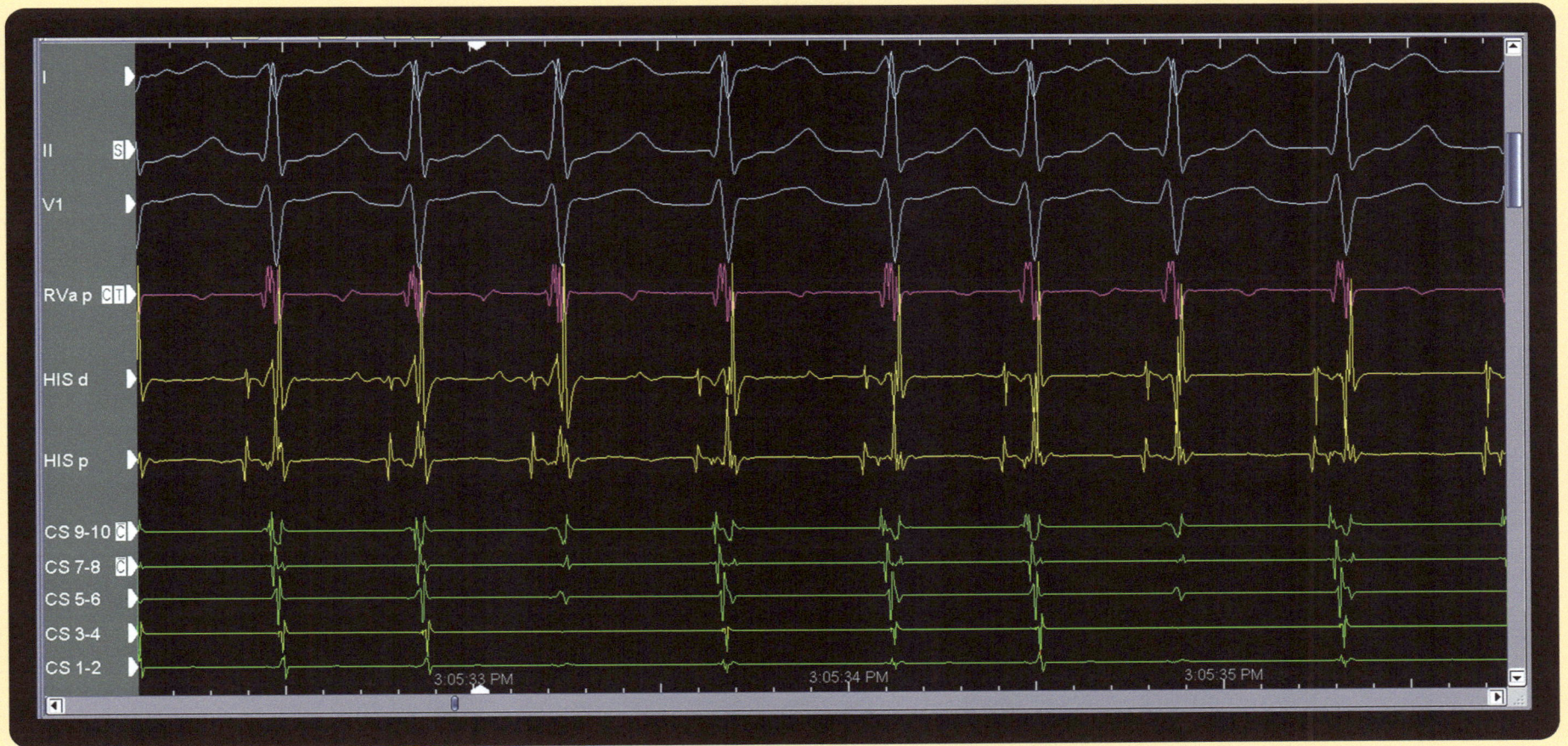

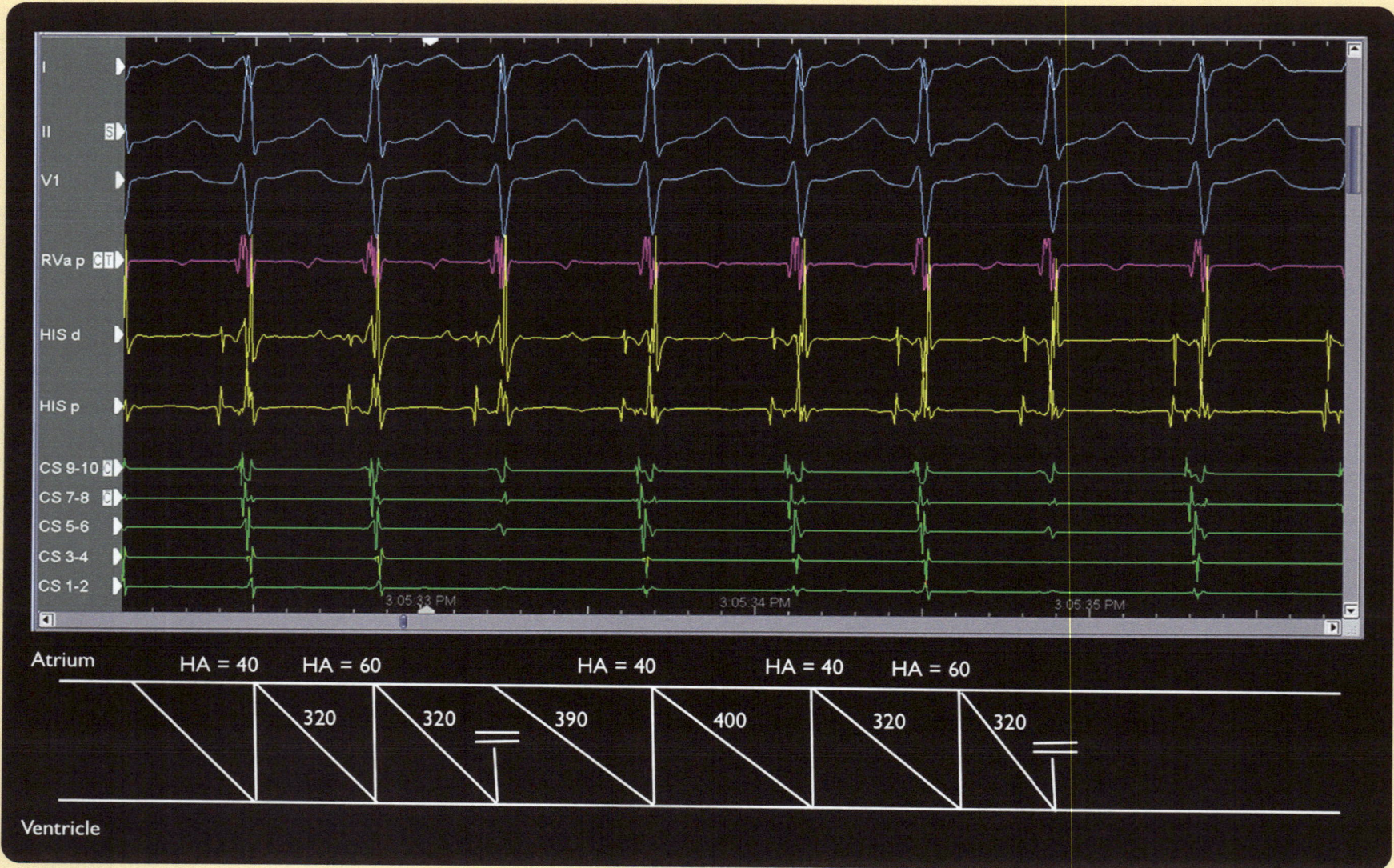

Atrium

HA = 40   HA = 60   HA = 40   HA = 40   HA = 60

320   320   390   400   320   320

Ventricle

Discussion

The previous tracings in sections 3 and 4 demonstrated block to the ventricle.

This tracing demonstrates block to the atrium proving that the AVNRT circuit does not require the atrium for maintenance of tachycardia. This observation occurs quite infrequently.

Unusual observation #2: Careful measurement reveals that there are 3 distinct A-H intervals (320, 390 and 400 ms). There are also 2 H-A intervals (40 and 60 ms).

These variable antegrade and retrograde pathways cause considerable "wobble" in the tachycardia cycle length (TCL).

2.7 Bundle Blocks

The tracing illustrates some important aspects about normal cardiac electrophysiology that is not limited to AVNRT alone. There are 3 interesting observations requiring explanation.

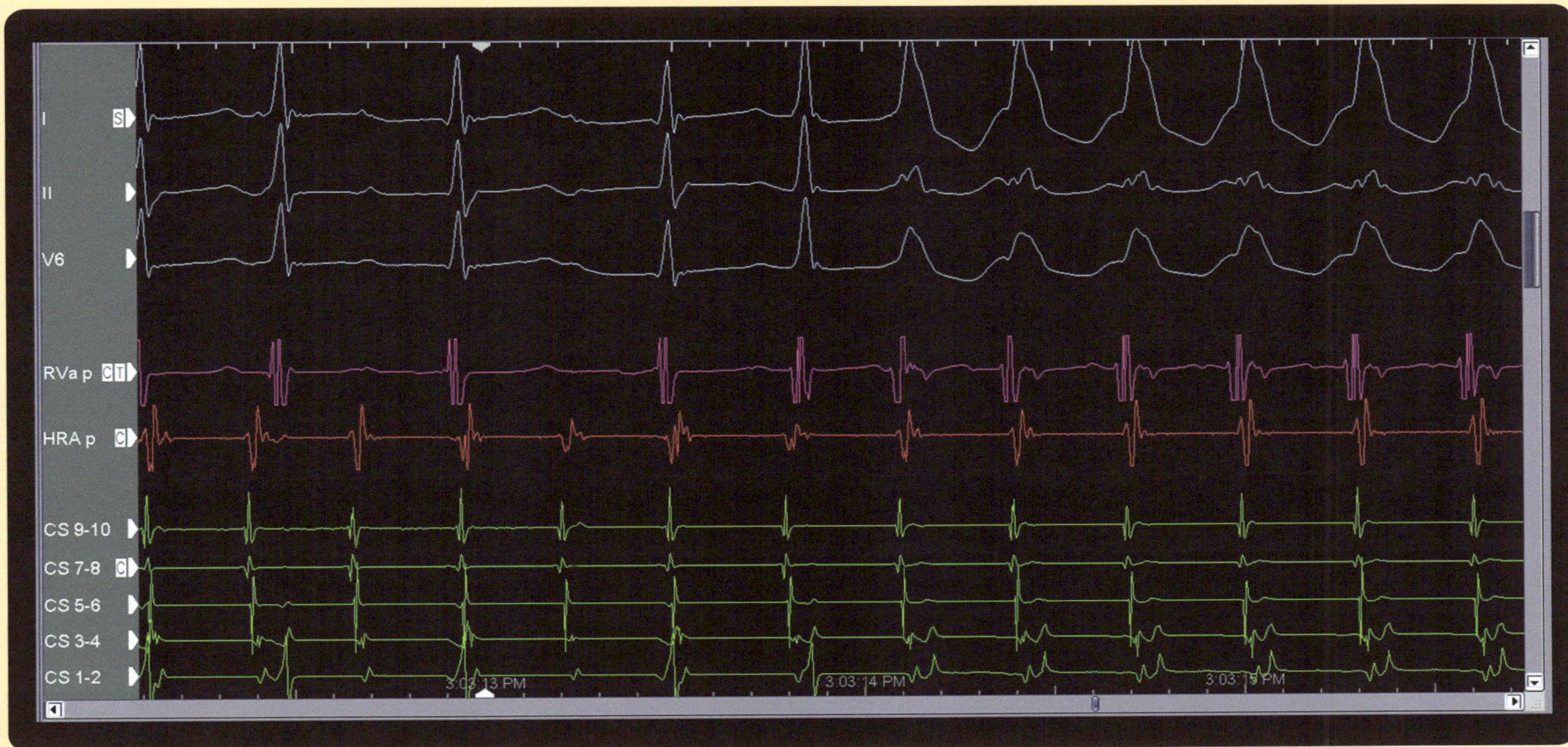

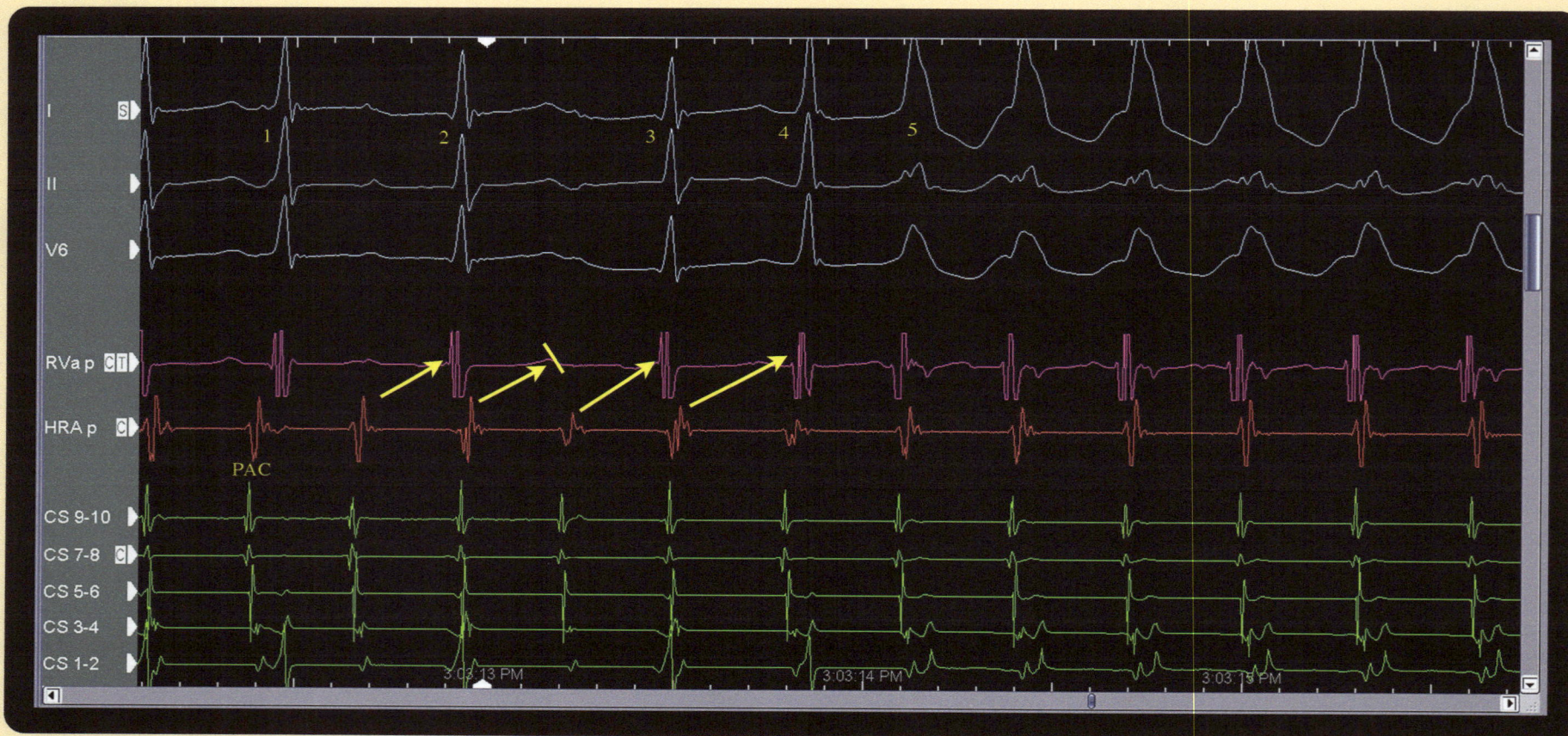

Discussion

The first 2 cycles show ongoing AVNRT with a 2:1 A-V relationship (similar to previous tracings).

At beat #4, 2 phenomena occur:

1.) Spontaneous resumption of 1:1 A-V relationship during ongoing AVNRT and, 2.) LBBB at beat #5.

Point 1: The most likely explanation for spontaneous resolution of the 2:1 A-V block during AVNRT is enhanced conduction velocity and shortening of the ERP of the His-Purkinje tissue below the AVNRT circuit (including the bundle branches), until, at beat #3, it was sufficiently short to allow resumption of conduction.

As mentioned in Section 4 of this chapter, this is sometimes referred to as "peel back."

Point 2: The occurrence of LBBB, once 1:1 A-V conduction had resumed, can be explained by failure of the left bundle branch system to sufficiently adapt (as the right bundle system had). An alternate explanation is that the LBBB is maintained by transseptal retrograde concealment into the left bundle branch system from the right bundle branch system.

Point 3: Beat #4 is interesting. Notice that the QRS morphology is subtly different from the QRSs on either side of it. This could be due to incomplete "peel back" of the right bundle branch system.

2.8 Echoes?

This is an exercise in methodically explaining the mechanism of each beat, one at a time. Differences in tachycardia mechanisms and their initiation can be quite subtle leading to misinterpretation.

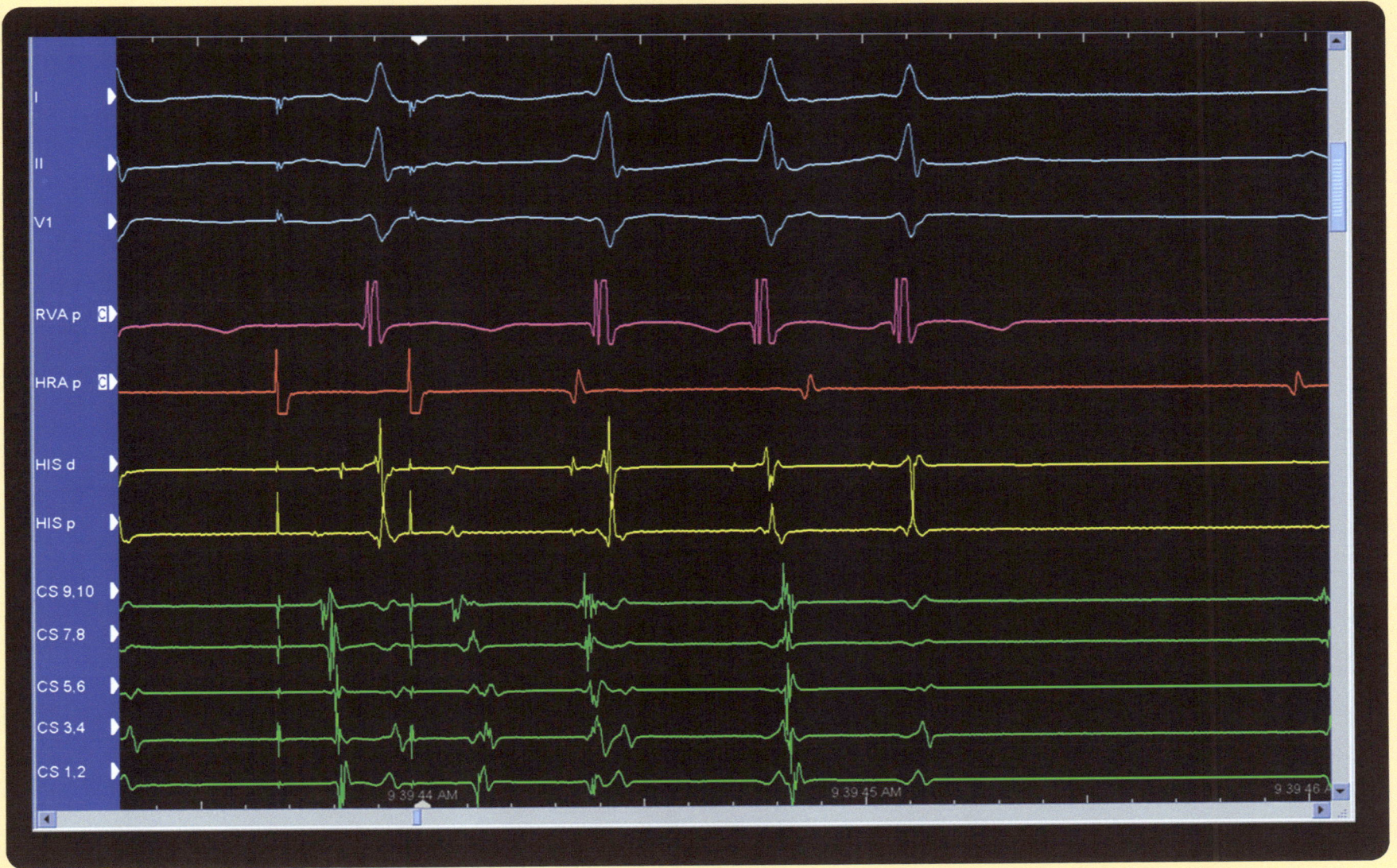

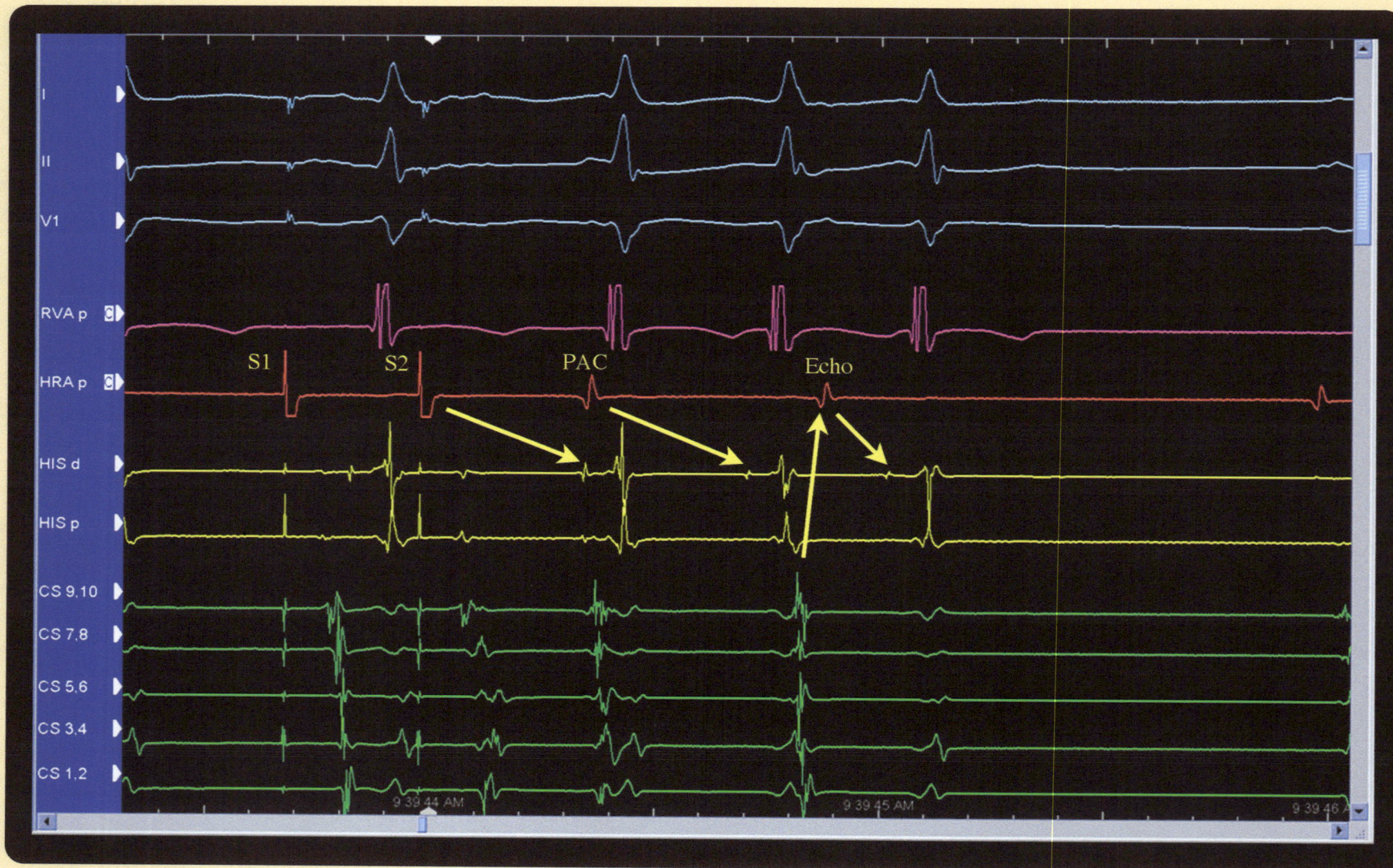

Discussion

The S2-initiated "A" conducted to the ventricles via the antegrade AV nodal slow pathway. The first arrow points to where the His deflection should have been. However, the His electrogram is partially obscured due to the occurrence of another early "A" on top of it. This must be a premature atrial contraction (PAC) since the earliest "A" appears in the HRA channel.

This PAC then conducts to the ventricles via an even slower antegrade AV nodal slow pathway. The wave then conducts back up a retrograde AV nodal fast pathway, producing an AV nodal echo beat. Subsequently, this echo beat conducted to the ventricles but failed to initiate AVNRT.

2.9 Atypical

What have we initiated and how was it initiated?

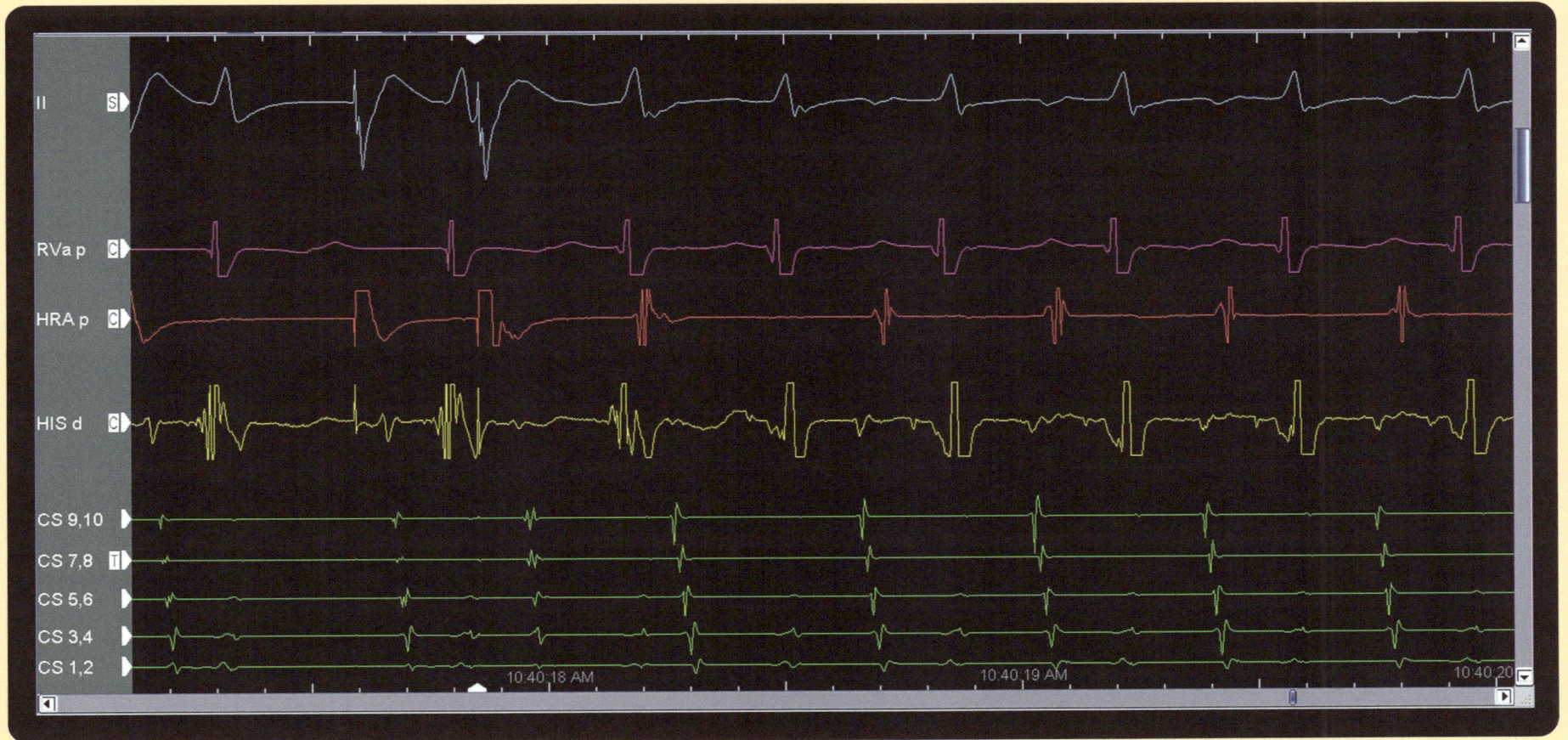

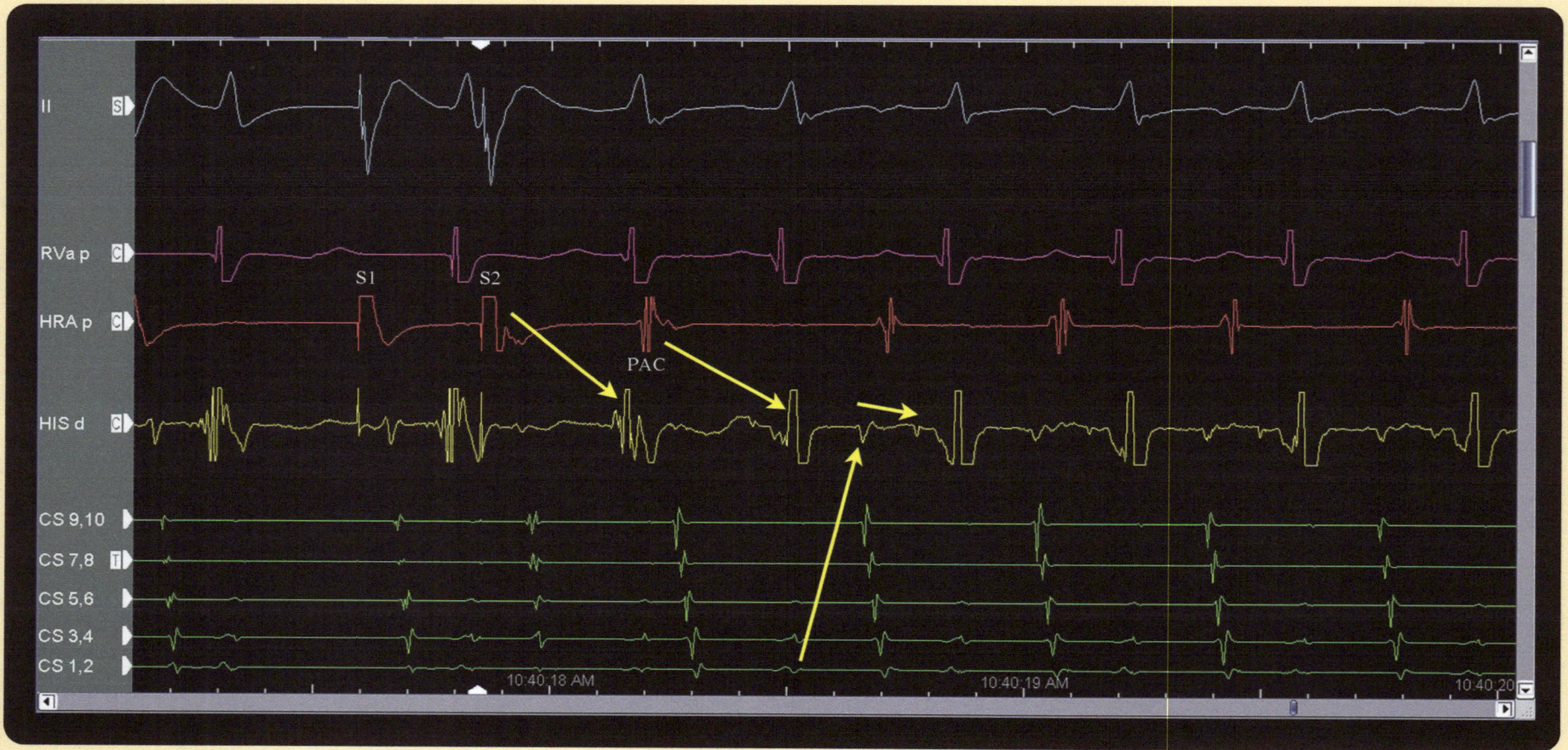

Discussion

What we have here is a long V-A tachycardia induced during atrial extra-stimulus testing. The differential diagnosis of a supraventricular tachycardia (SVT) with a 1:1 AV relationship and a long V-A interval include: AVRT, atrial tachycardia (AT), or atypical AVNRT. Let's examine the initiation.

The S2-induced "A" conducts to the ventricles in the usual manner. Unfortunately, the His deflection is temporarily missing.

Following the S2, we see a PAC (earliest "A" at the HRA). It conducts to the ventricles via an antegrade AV nodal slow pathway followed by another "A" which is earliest at the His channel. This sequence then repeats itself.

The most likely explanation is that the "A" conducted antegradely down the AV nodal slow pathway, then conducted retrogradely back up another retrograde AV nodal slow pathway (long V-A interval). On subsequent beats, the antegrade limb of the circuit is an AV nodal fast pathway.

This is an example of atypical AVNRT...sometimes described as "down the fast and up a slow" AVNRT.

In the differential diagnosis of long V-A tachycardias, AVRT must always be considered. The retrograde limb could be a midseptal accessory pathway.

As well, a rather slow "low-atrial" automatic focus resulting in atrial tachycardia is also on the differential, but less likely.

2.10 Will S3 help?

When AVNRT cannot be initiated by standard S1S2 extra-stimulus pacing, it is often useful to add in an S3. However, one should always question whether the S3 will help or hinder!

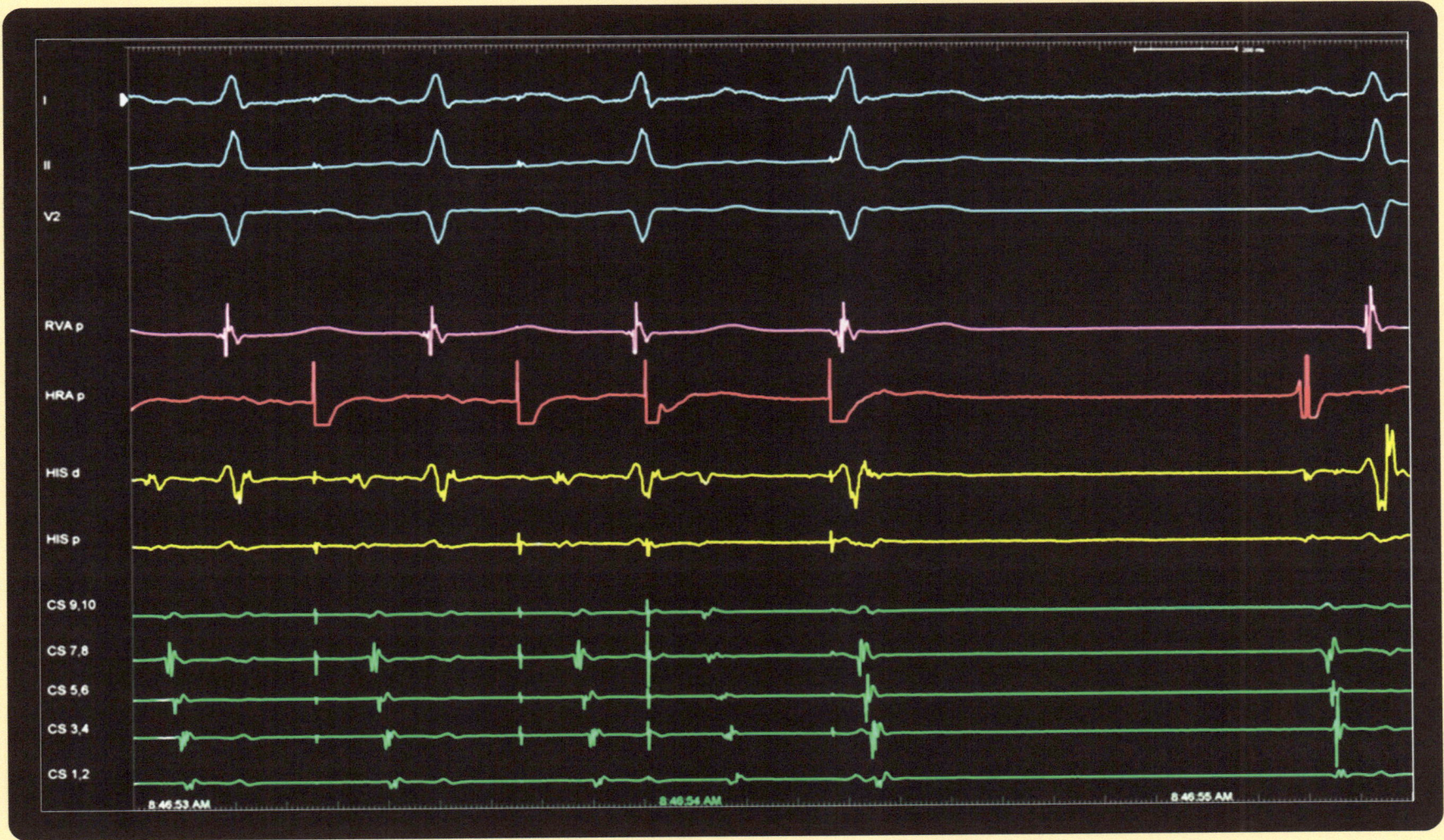

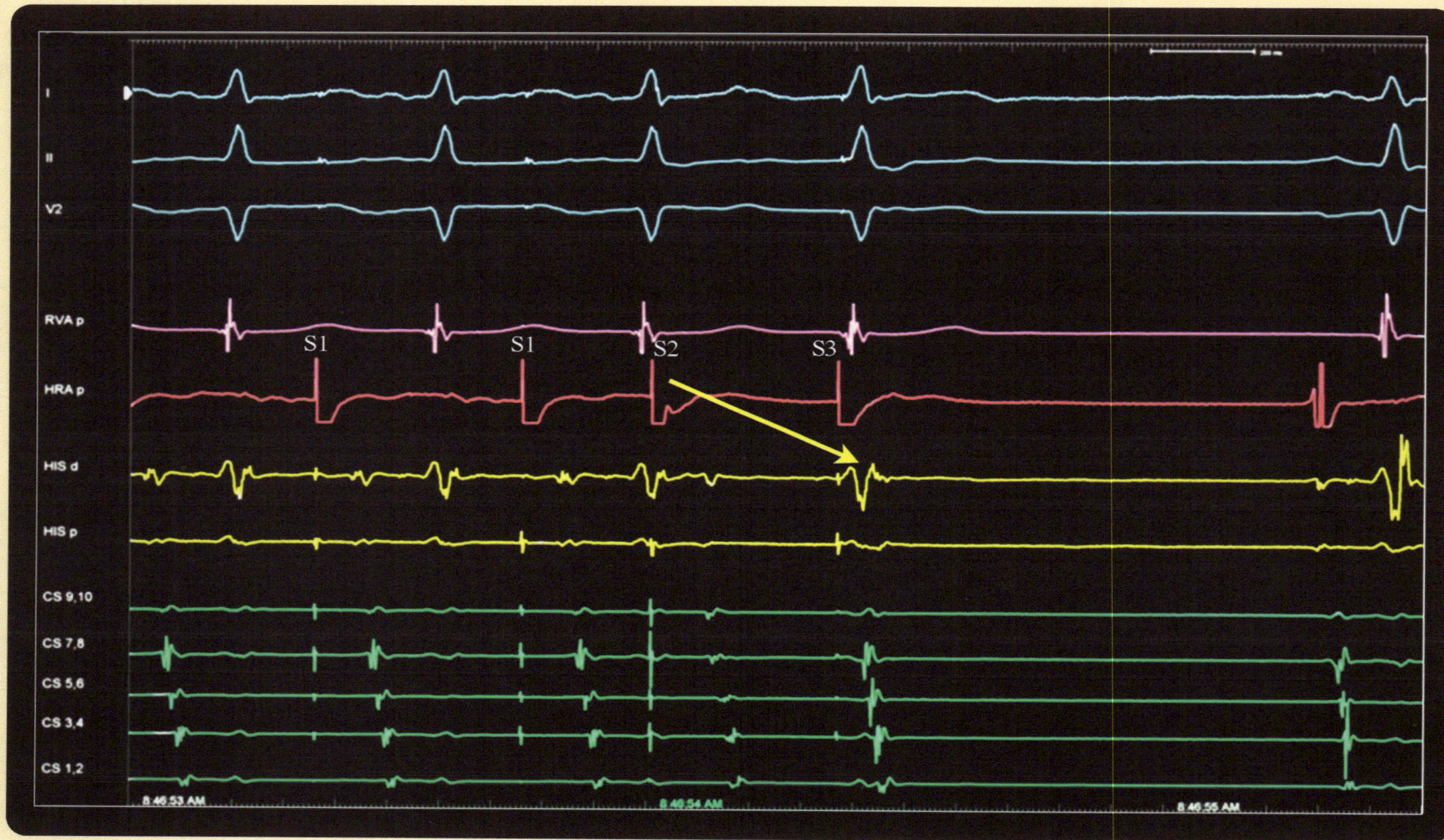

Discussion

Earlier in this EP study, we demonstrated single AV nodal echo beats initiated by atrial S2's but no tachycardia was inducible. So an S3 was added. There are 3 possible effects of the S3:

1. The S3-induced "A" wavefront could conduct over an even slower antegrade AV nodal pathway and initiate AVNRT.

2. The S3-induced "A" wavefront could conceal antegradely into the AV nodal fast pathway, rendering it refractory and thus preventing any retrograde fast pathway echo beats.

3. The S3-induced "A" wavefront renders the atrial myocardium adjacent to the retrograde limb of the circuit refractory thus blocking the initiation of AVNRT. This option is functionally the same as option #2 above.

When we removed the S3, the single AV nodal echo beat returned thus proving option#2 above. The S3 must have antegradely penetrated the retrograde limb of the AVNRT circuit rendering it refractory and thus blocking the echo.

2.11 PAC Effect

On the left side of the tracing, we have ongoing typical AVNRT. However, junctional tachycardia remains on the differential diagnosis. Can we differentiate between these two possibilities? Perturbing the tachycardia by introducing a PAC may provide the answer. This is what is illustrated in this tracing.

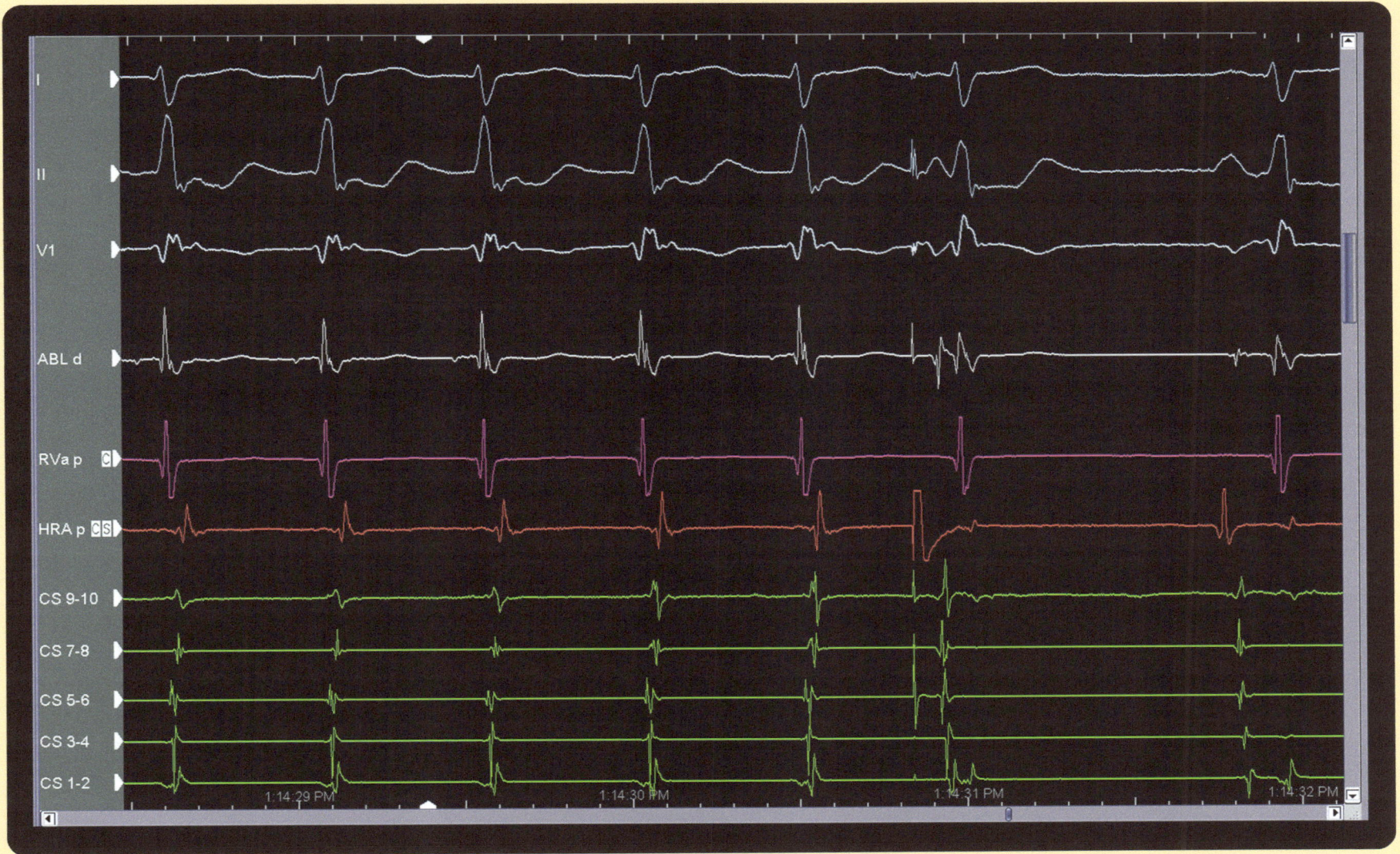

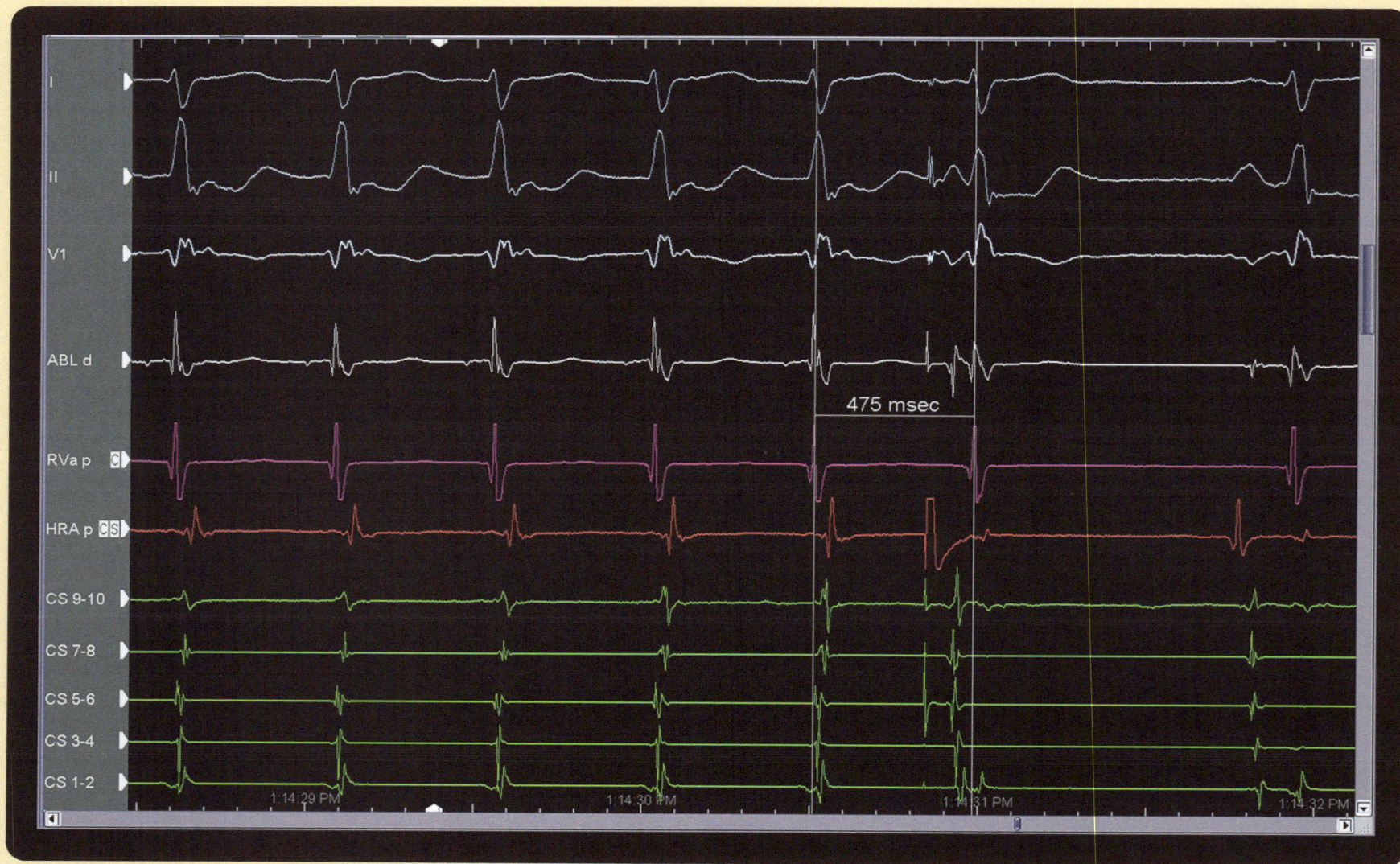

475 msec

Discussion

The tachycardia terminated simultaneous with delivery of the single paced "A" (PAC equivalent). Measuring the V-V intervals on the right ventricular apex (RVA) channel before and after the PAC, reveals that the paced "PAC" terminated the tachycardia *without* advancing the next "V"! There are 2 possible explanations for this:

Scenario 1: This is junctional rhythm (not AVNRT). If this is true, the paced "PAC" would be expected to overdrive the junctional rhythm and, in doing so, advance the next "V," but it did not.

Scenario 2: This is truly AVNRT. If this is true, we would expect that a paced "PAC" could antegradely penetrate the retrograde limb of the circuit, rendering it refractory and therefore break the tachycardia without advancing the next "V." This is the identical concept we illustrated in the previous tracing.

This tracing is most compatible with Scenario #2.

Notes

Cardiac Electrophysiology 2
CHAPTER 3   AVRT

3.1 Subtle

Changes during pacing maneuvers can be quite subtle, requiring focus and concentration. Explain the atrial activation sequence associated with each paced ventricular beat.

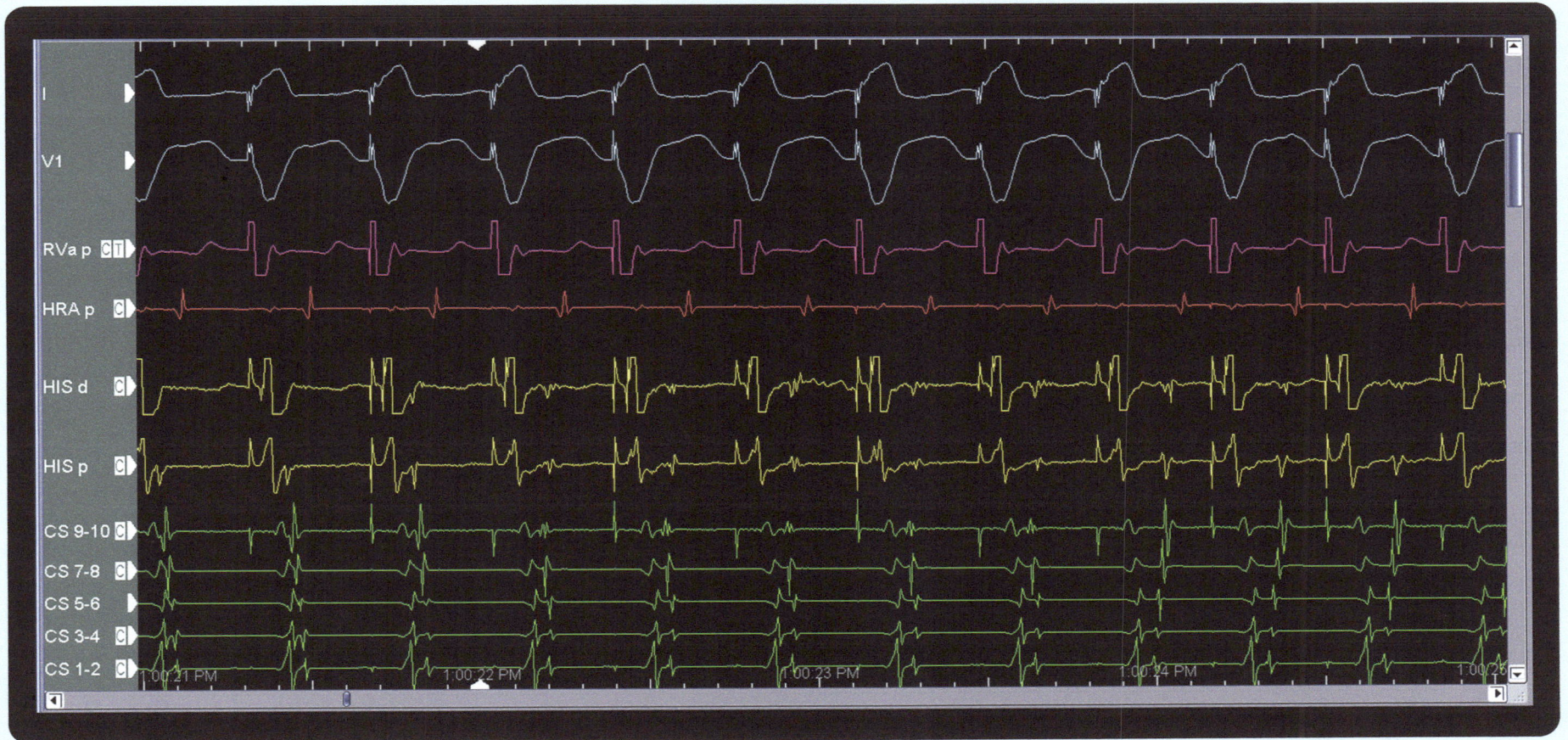

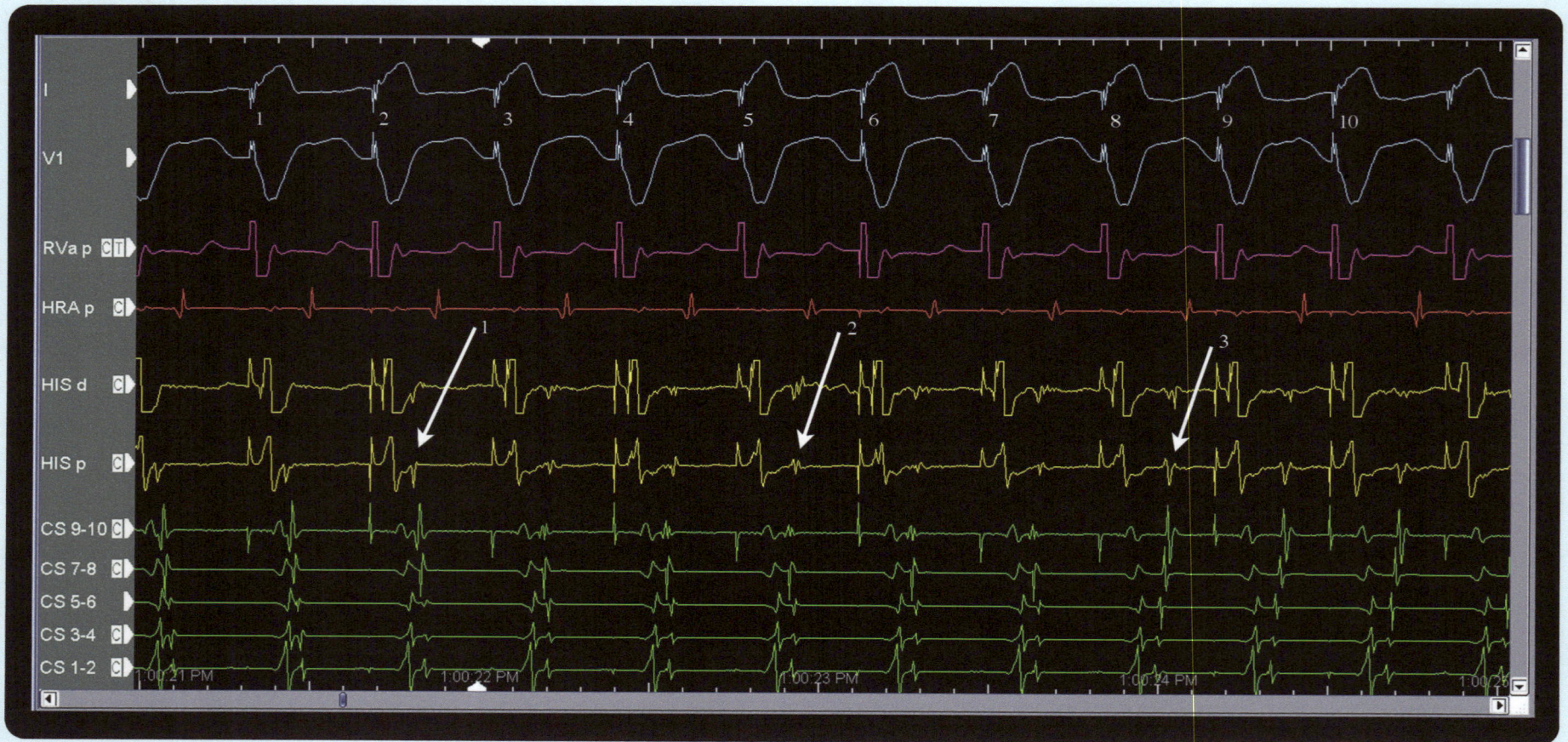

Discussion

QRS #1 and #2 have a "central" atrial activation sequence (His "A" being the first atrial activation). This is most compatible with conduction over the normal A-V conduction system but also compatible with conduction over a septal accessory pathway.

QRS #3, 4, 5, 6, and 7 demonstrate a subtle change. The His "A" now follows the proximal CS 9-10. Since CS 9-10 has not moved and is now second in the atrial activation sequence this suggests conduction over a posteroseptal AP although conduction over a "slow" AV nodal pathway is also possible.

In QRS #8, 9, 10, the distal CS 1-2 is now earliest ("eccentric" atrial activation) indicating conduction over a left lateral AP.

This patient did indeed have 2 accessory pathways, a posteroseptal and a left lateral.

3.2 Pacing Site

Multiple pacing sites often enhance the probability of inducing tachycardia when the standard pacing site is not productive. Explain the rationale.

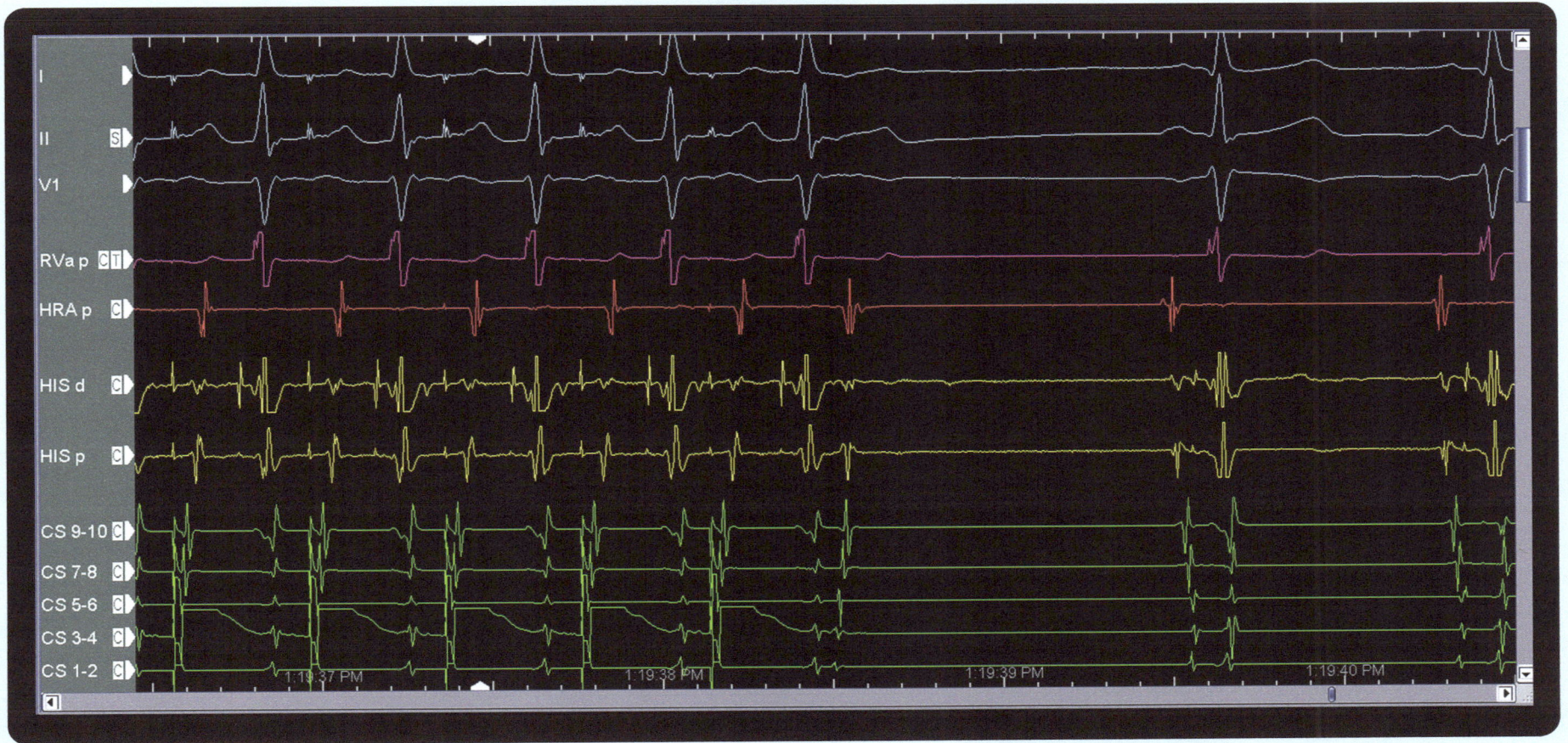

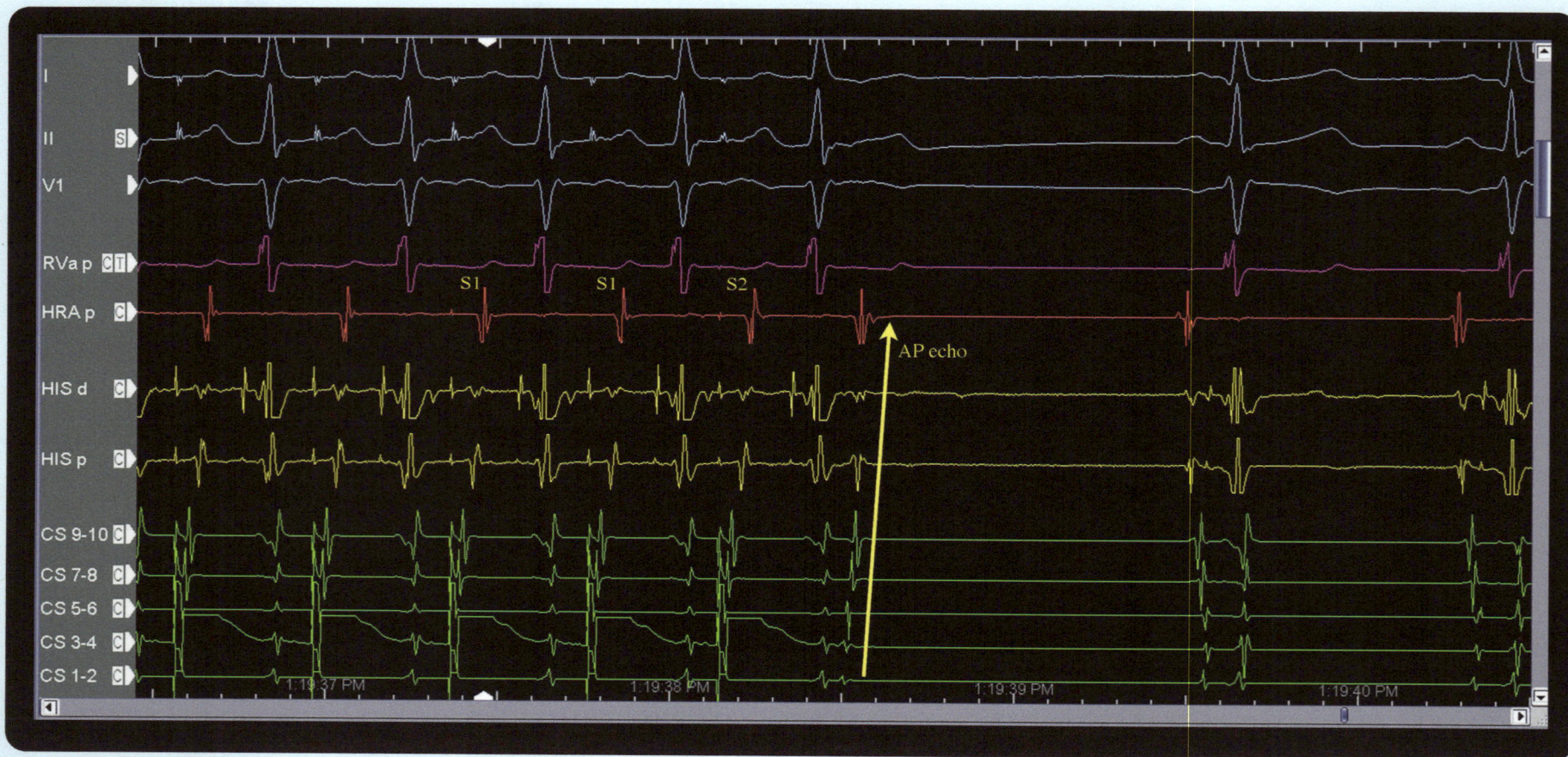

Discussion

During atrial extra-stimulus pacing from the HRA, there was no evidence of preexcitation. This can be due to:

1. The AP conducts retrograde only.

2. Right atrial pacing favors AV nodal conduction over that of a distant left lateral AP. A normal, short A-V conduction time may result in the ventricles being depolarized before the atrial impulse can get to the AP. Ventricular pacing (not shown here) confirmed a left lateral AP.

There were no AP echoes during HRA extra-stimulus testing. However, we did induce AP echoes while pacing from CS 3-4. Why?

Pacing from the HRA depolarizes the lateral left atrium relatively late, such that a wave of depolarization trying to return to the left atrium via a left lateral AP will find the atrial insertion site refractory.

However, pacing the atrium closer to the atrial insertion site of the AP (in this case CS 3-4) allows the atrial insertion site more time to recover excitability. This allows the retrograde wave of depolarization coming from the AP to depolarization of the left atrium generating an echo beat.

3.3 AP Revealed

The term "concealed conduction" is used since one does not
see this conduction directly on the ECG or EGM signals
but infers that it must be there by other observations that are
difficult to explain otherwise. This is an example of
concealed retrograde penetration of the AV node as an
explanation for the ECG observations.

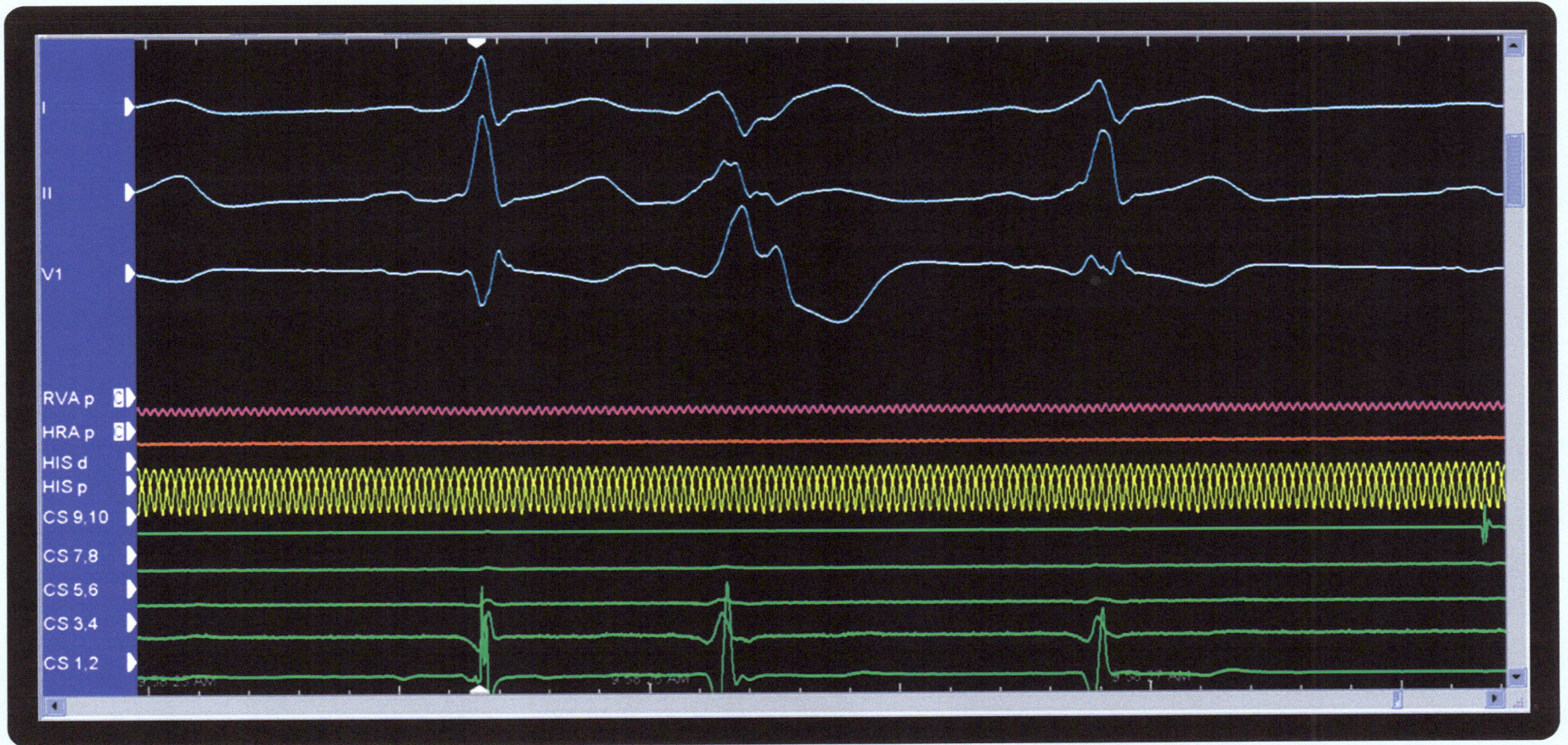

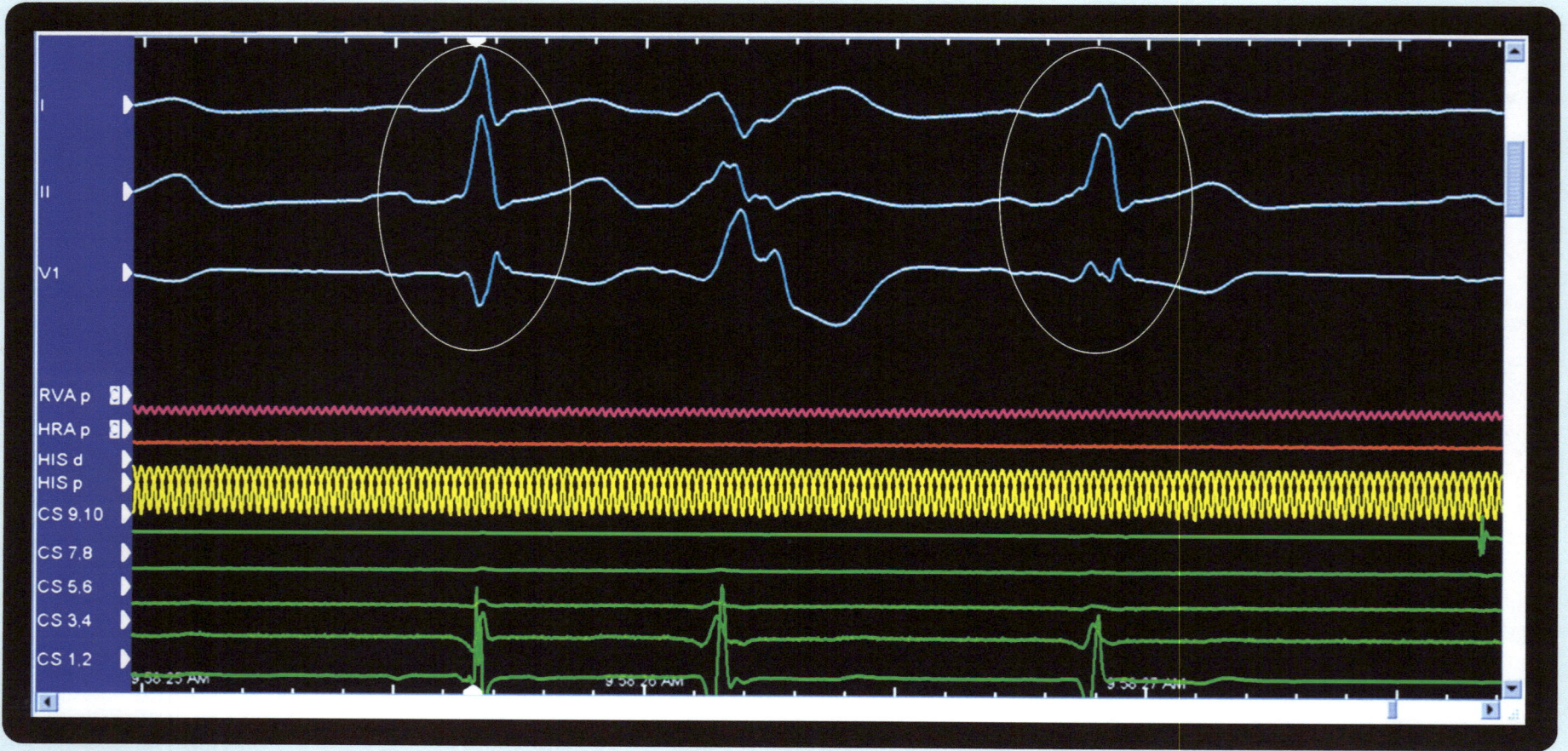

Discussion

The first beat is slightly preexcited with a delta wave best appreciated in leads I and II. This suggests that the ventricle is being depolarized initially over an AP (i.e., delta wave) and then predominately over the normal A-V conduction system.

The second cycle is a premature ventricular contraction (PVC) created by catheter manipulation.

The third cycle is again different, a little broader and "more" preexcited. This is compatible with more of the ventricle being activated by the accessory pathway and less by the normal A-V conduction system.

This is most compatible with "concealed" retrograde conduction into the AV node by the PVC, which impedes or blocks AV nodal conduction and thus allows the AP to predominately depolarize the ventricle.

3.4 ERPs

There are several key observations in this tracing.

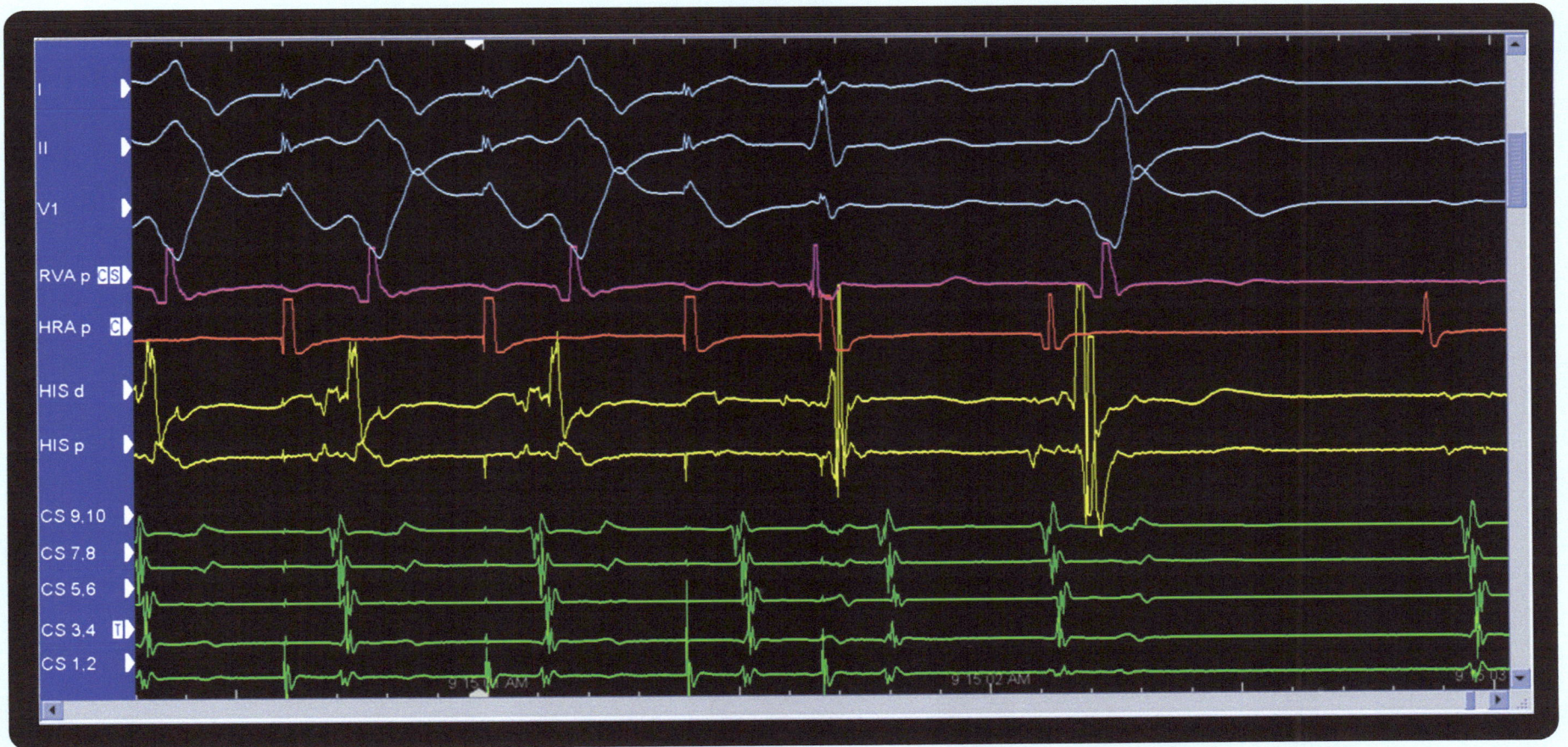

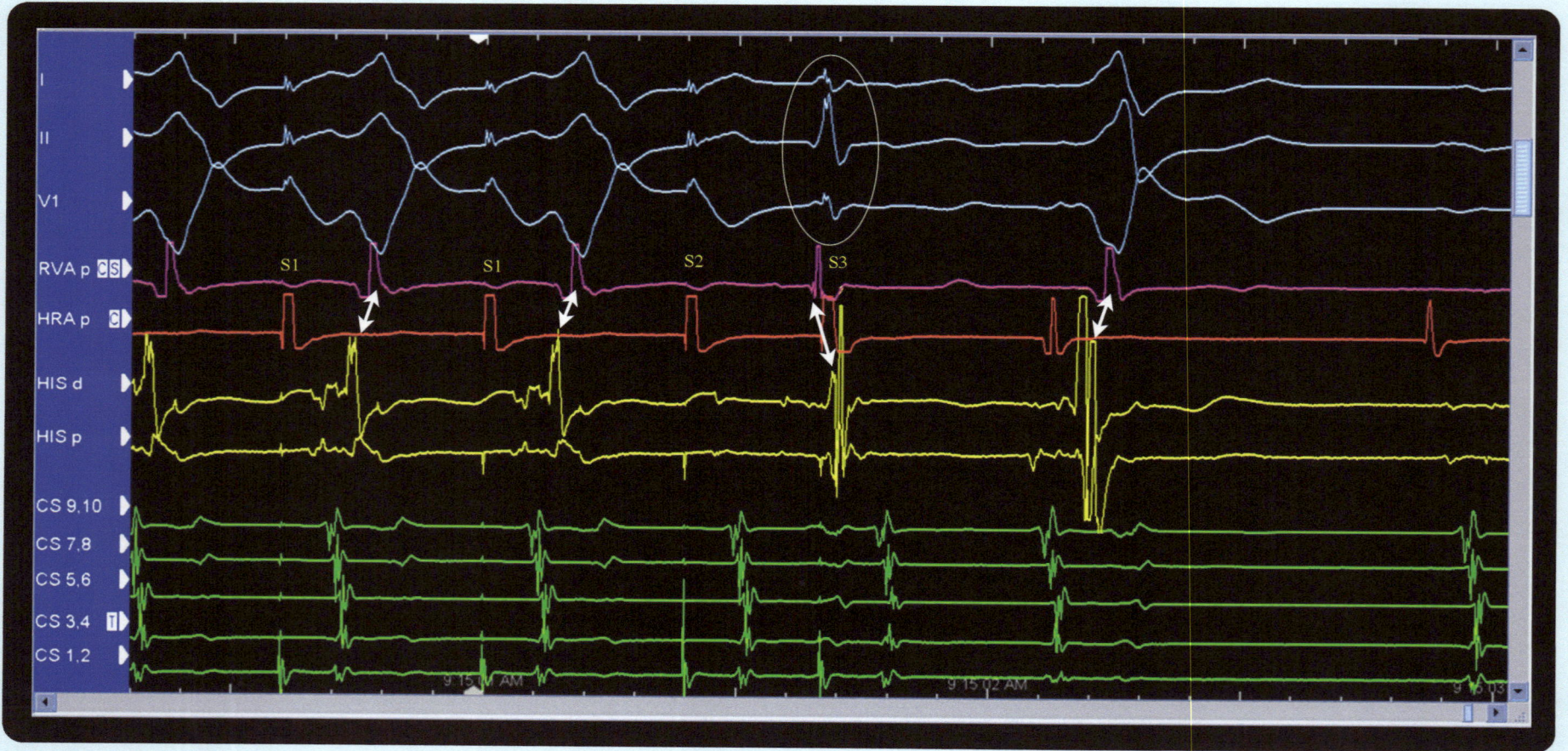

Discussion

1. *Arrows 1 and 2* are highlighting the relationship between the RVA "V" and the His "V." With normal conduction over the AV node and RBBB, the RVA is activated well before the His region "V." The basal His "V" is earlier here. The His bundle electrogram is not clearly seen but, if present, must be buried within the QRS. This is diagnostic of conduction over an accessory pathway relatively close to the His bundle catheter and the surface ECG shows marked preexcitation with a LBBB pattern indicative of a right-sided AP.

2. The S2 has reached the ERP of the AP and thus blocks in the AP. The wave of depolarization conducts exclusively over the AV node generating a narrow QRS. Notice that the His "V" and the RVA "V" relationship has reversed. (*Longer white arrow* #3.)

3. The S3 captured the atrium with a long pause before the next QRS. One possibility is that the S3 blocked in both the AV node and the AP and therefore the next QRS is an atrial escape cycle (His "A" before HRA) with a fully preexcited QRS.

Alternatively, it is possible that the S3 conducted over a very slow AV nodal slow pathway and that the cycle is an AV node echo cycle with "bystander" preexcitation. Slow AV node pathways often coexist with accessory pathways. Since this observation was not reproducible, it is unlikely that this possibility is correct.

3.5 VA?

An interesting question during ventricular pacing with eccentric atrial activation via an accessory pathway is: "Is there V-A conduction over the AV node?" Does this tracing help determine if retrograde conduction over the normal A-V conduction system will be present following successful ablation of the AP?

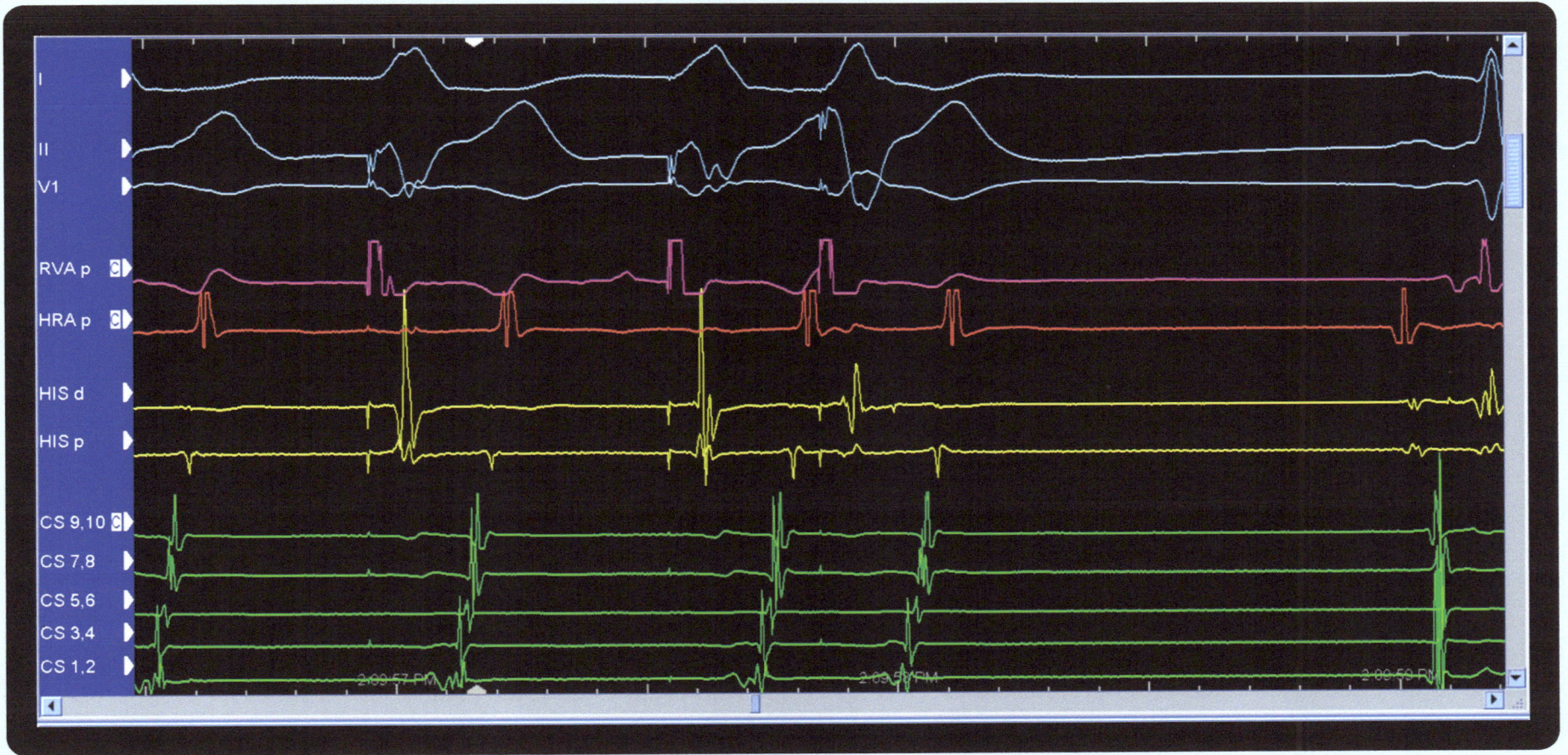

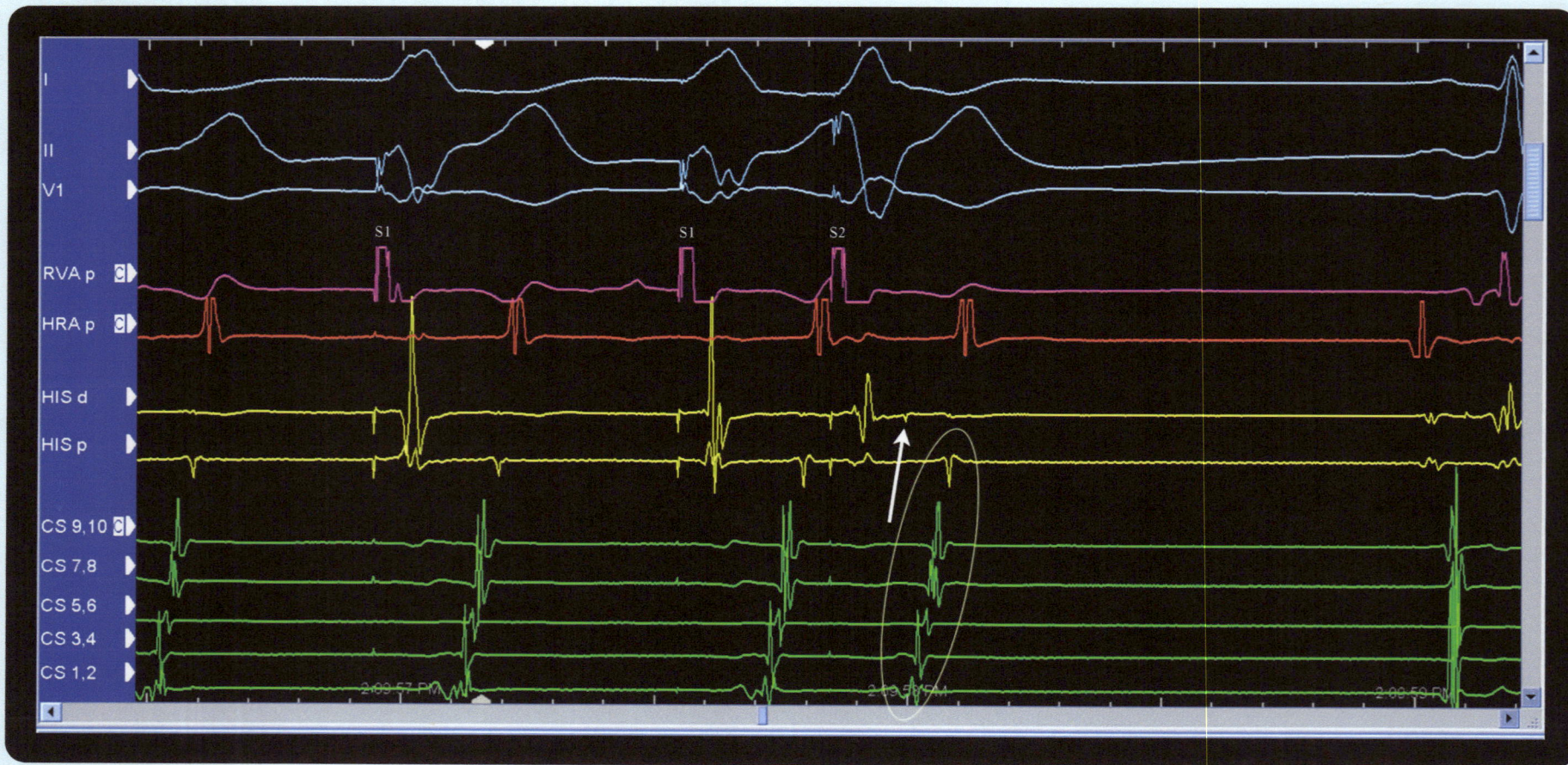

Discussion

With the ventricular extra-stimulus (S2), atrial activation is eccentric with earliest activity in the distal CS 1-2. A retrograde His deflection (*arrow*) is observed indicating that conduction reached the level of the AV node but there is no suggestion that conduction reached the atrium thereafter. A retrograde His is indication that the S2 reached the ERP of the right bundle branch. The wave of depolarization must therefore traverse the interventricular septum and arrive at the AV node via the left bundle system. This increased delay (due to transseptal conduction) in the arrival of the wave of depolarization at the AV node, should have allowed AV nodal conduction to occur.

The absence of retrograde conduction through the AV node may have several explanations:

1. There is no retrograde V-A conduction over the AV node at all. This would be consistent with this example.

2. There is retrograde V-A conduction but the conduction time over the AV node is longer than the conduction time over the accessory pathway and hence not seen.

3. There is retrograde conduction over the AV node but it is not present at the pacing cycle length tested in this example.

4. The AV node may have been contused ("bumped") during insertion of the catheters and will recover subsequently.

In this patient, there was no V-A conduction following successful ablation of the AP.

3.6 ERPs?

This tracing indicates 3 levels of block! Can you find them?

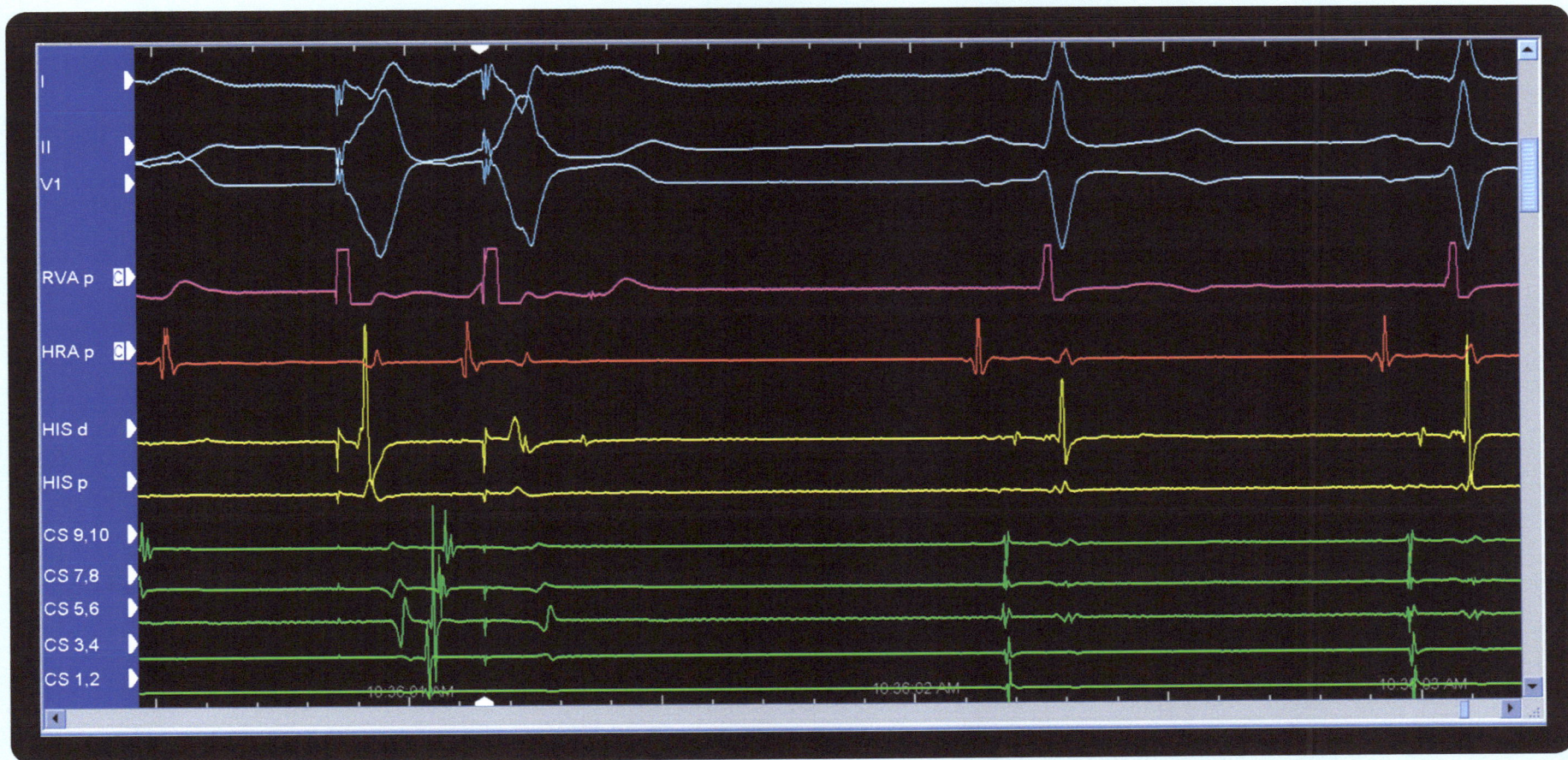

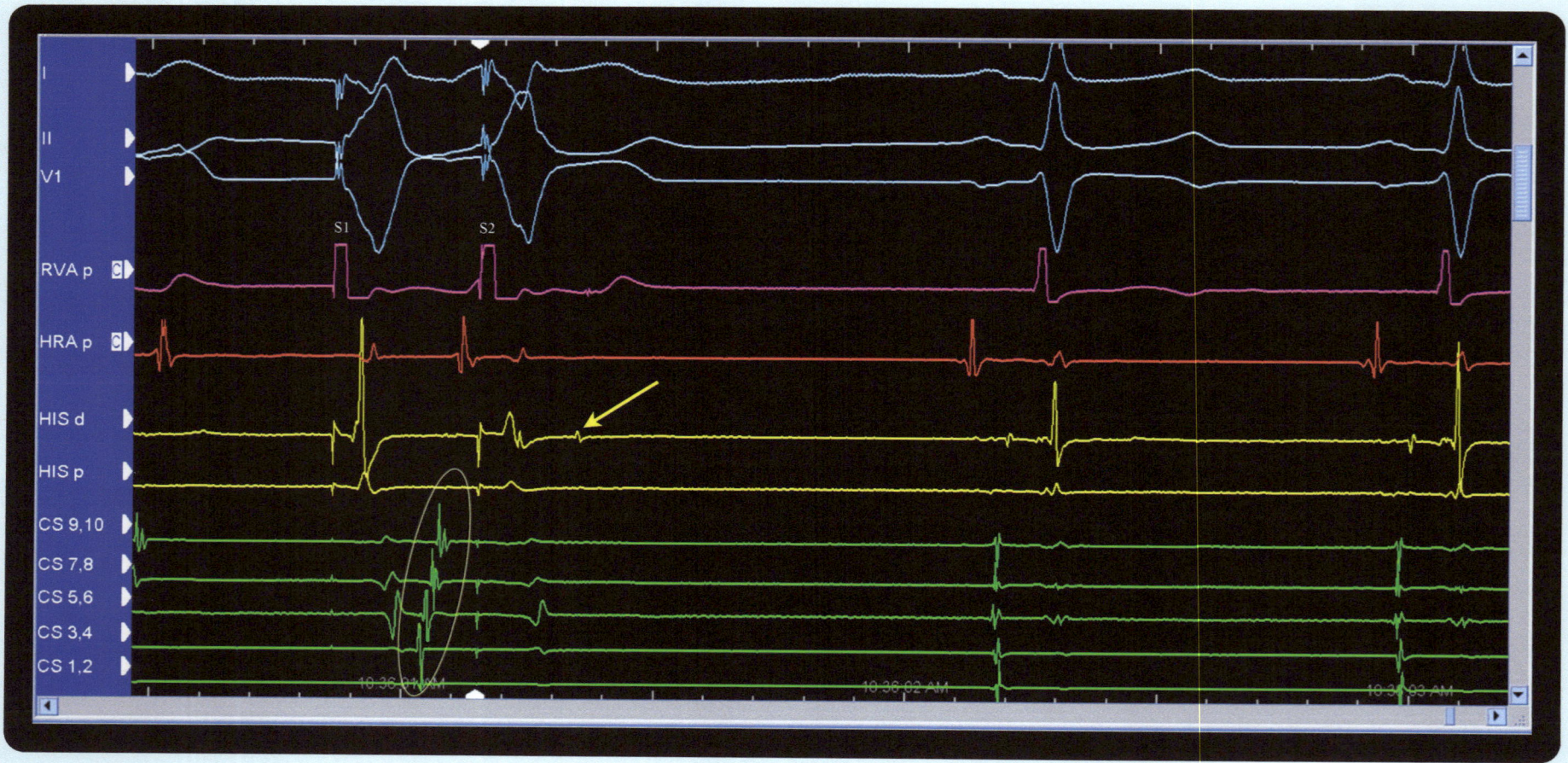

Discussion

Ventricular depolarization following the S1 demonstrates eccentric retrograde atrial conduction (distal CS 1-2 early) indicating conduction over a left lateral AP. There is no atrial activity following the ventricular extra-stimulus (S2). This is indication of conduction block over the AP. That's one.

In addition, this also means that there is no retrograde conduction over the normal AV conduction system. That's two.

Finally, we see a retrograde His (*arrow*) well after the ventricular electrogram (VEGM) on the distal His channel. The retrograde His usually appears very early at the onset of ventricular depolarization. This marked delay suggests that retrograde block in the right bundle branch occurred, necessitating transseptal conduction and entry to the His via the left bundle branch. That's three.

3.7 His Pacing

This is an example of parahisian pacing. It is performed to establish the presence or absence of a septal accessory pathway. The catheter is positioned in the His bundle region and advanced toward the ventricle to minimize any atrial electrogram (AEGM) and thus avoid atrial capture. What is your interpretation of this pacing maneuver?

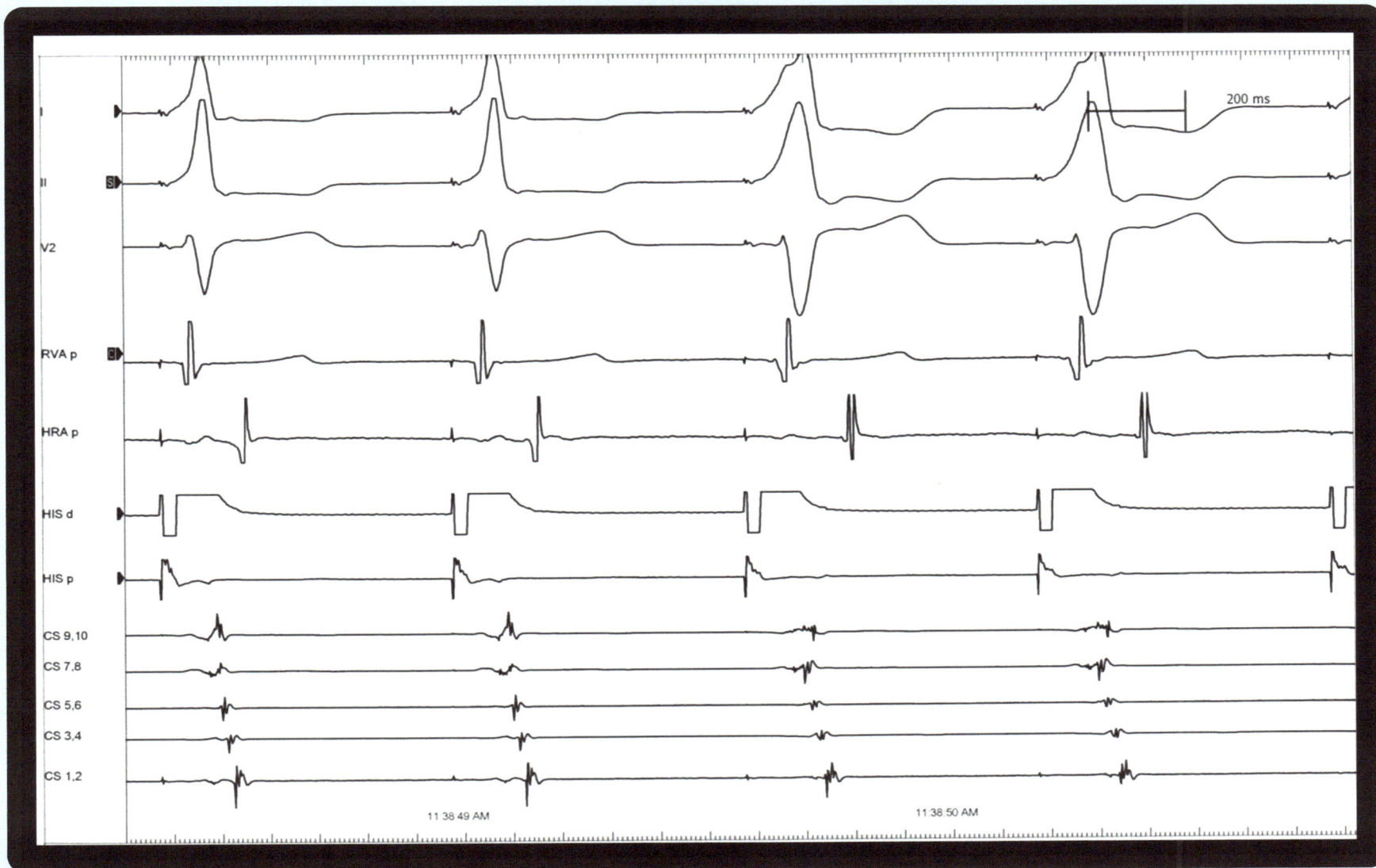

200 ms

I

II

V2

RVA p

HRA p

HIS d

HIS p

CS 9,10

CS 7,8

CS 5,6

CS 3,4

CS 1,2

11:38:49 AM   11:38:50 AM

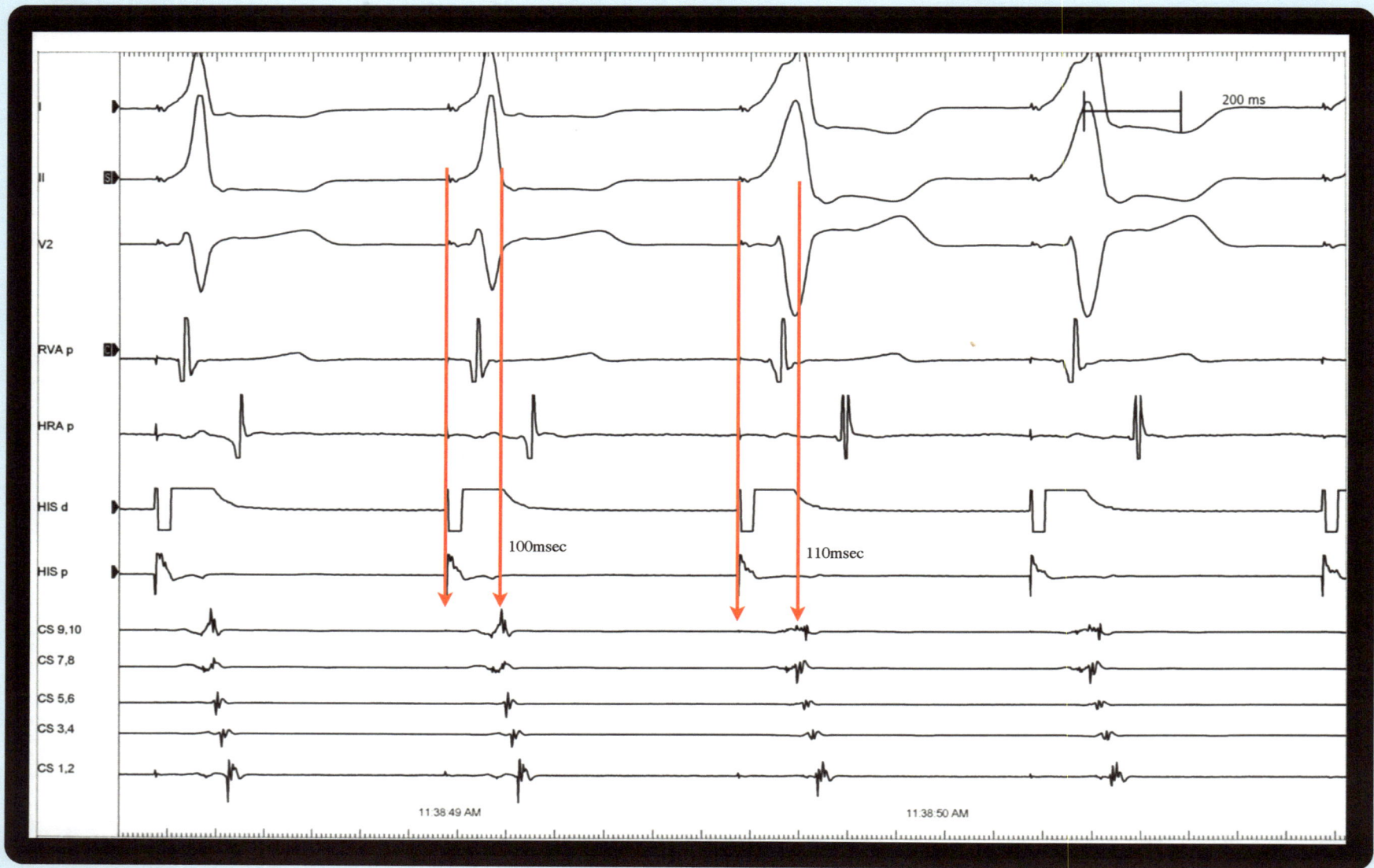

200 ms

I

II

V2

RVA p

HRA p

HIS d

HIS p

100msec

110msec

CS 9,10

CS 7,8

CS 5,6

CS 3,4

CS 1,2

11:38:49 AM

11:38:50 AM

Discussion

The first 2 beats display a narrow-complex QRS that is virtually but not entirely normal. This suggests capture of the His bundle *and* the ventricular myocardium adjacent to it simultaneously. The stimulator output was 20 mA and 2 ms wide. Take note of the Stim-A time.

The third and fourth beats are wide QRS complexes indicating loss of His bundle capture and capture of local myocardium only. The stimulator output was 2 mA and 2 ms wide. Again, take note of the Stim-A time.

The difference in Stim-A time between the wide-complex QRSs and the narrow-complex QRSs is 10 ms. This indicates the presence of a septal AP.

A difference in Stim-A time greater than 50 ms indicates the *absence* of a septal AP at the cycle length tested. It is not, strictly speaking, a "nodal" response, as commonly described, since it can be seen with accessory pathways distant from the septal region.

3.8 Narrow QRS

This patient has Wolff-Parkinson-White (WPW) with a consistently preexcited ECG. Why is the second cycle not preexcited, that is, demonstrating a normal QRS?

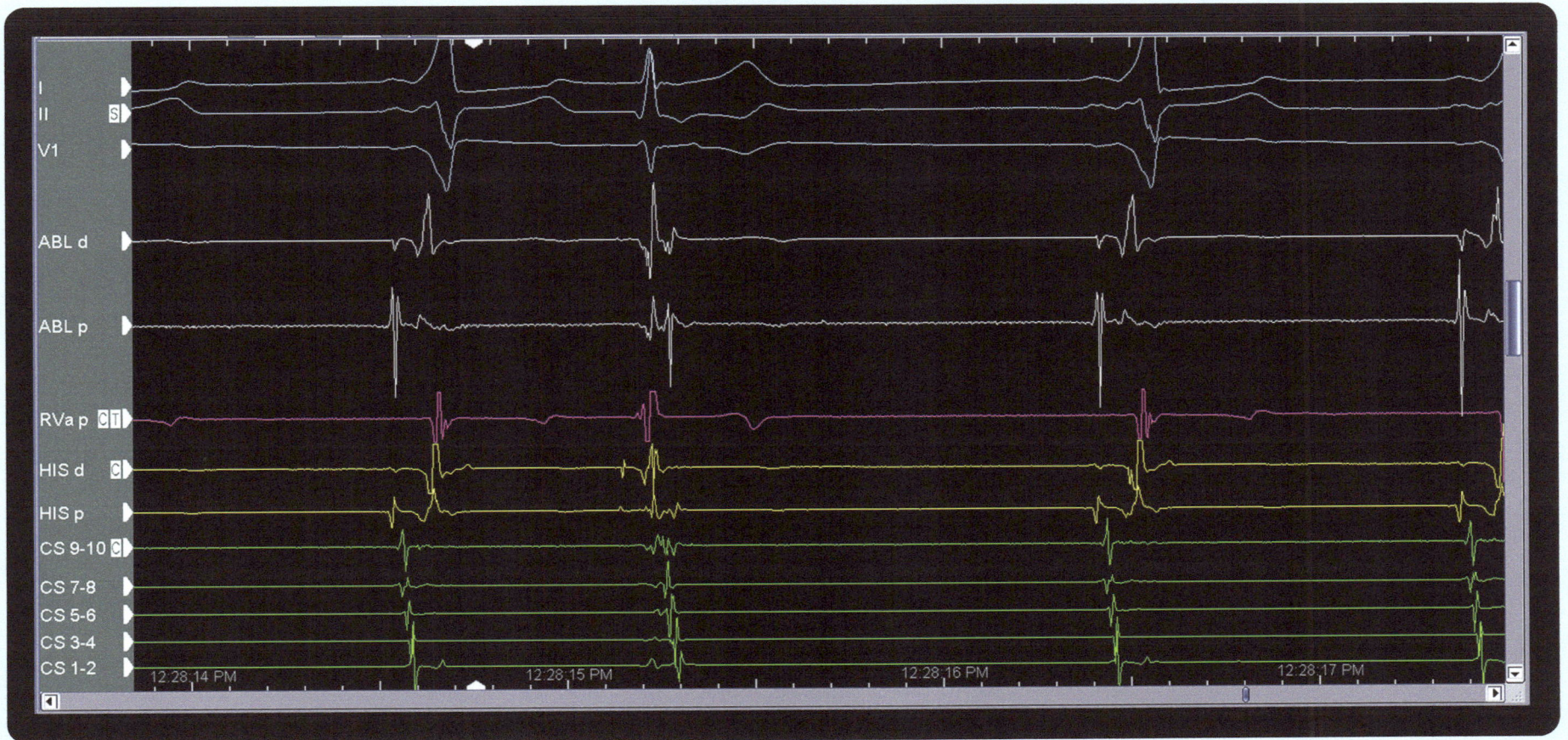

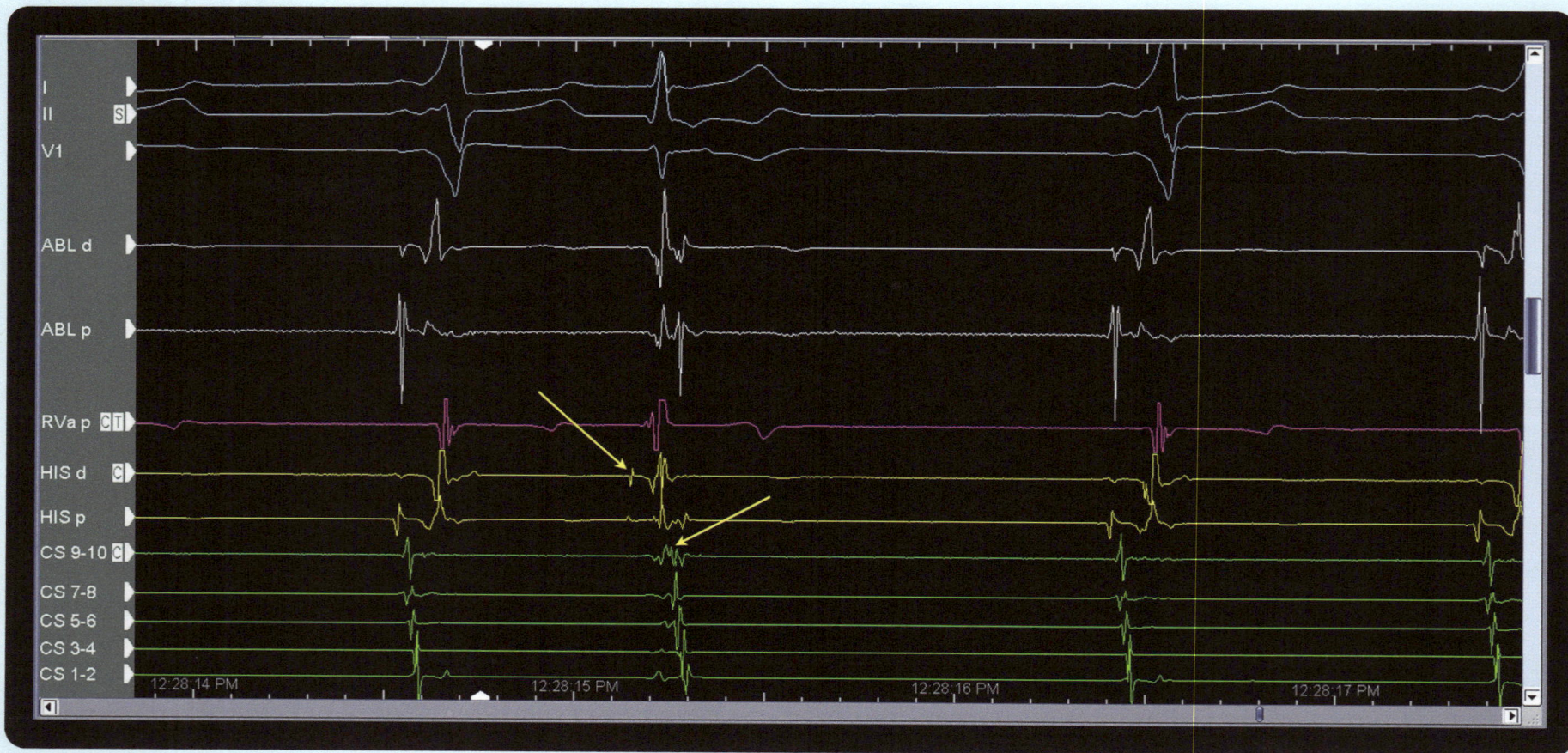

Discussion

The first intracardiac signal in the narrow second beat is a His deflection, indicating an extrasystole from the His region. This is commonly observed during His catheter placement. Since the junctional focus is below the insertion site of the AP, there will be no preexcitation.

The junctional extrasystole in general may conduct to the atrium over: 1) either; 2) both; or 3) neither of the AV node and the AP. In this instance, there is retrograde conduction but there is not enough information in this tracing to make this distinction.

3.9 Narrow QRS 2

This tracing shows preexcitation on beat #1, but the subsequent beats are not preexcited. Why?

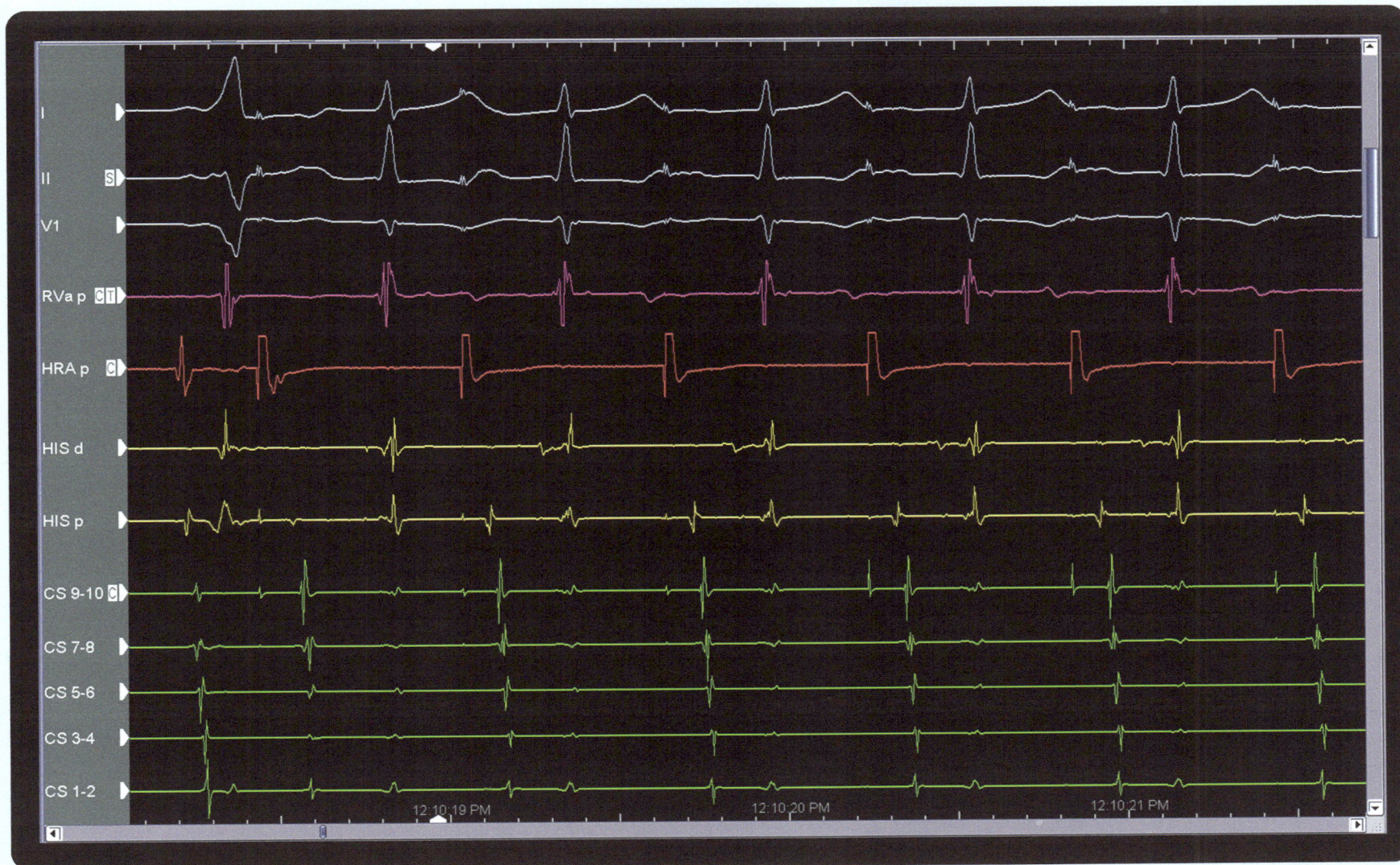

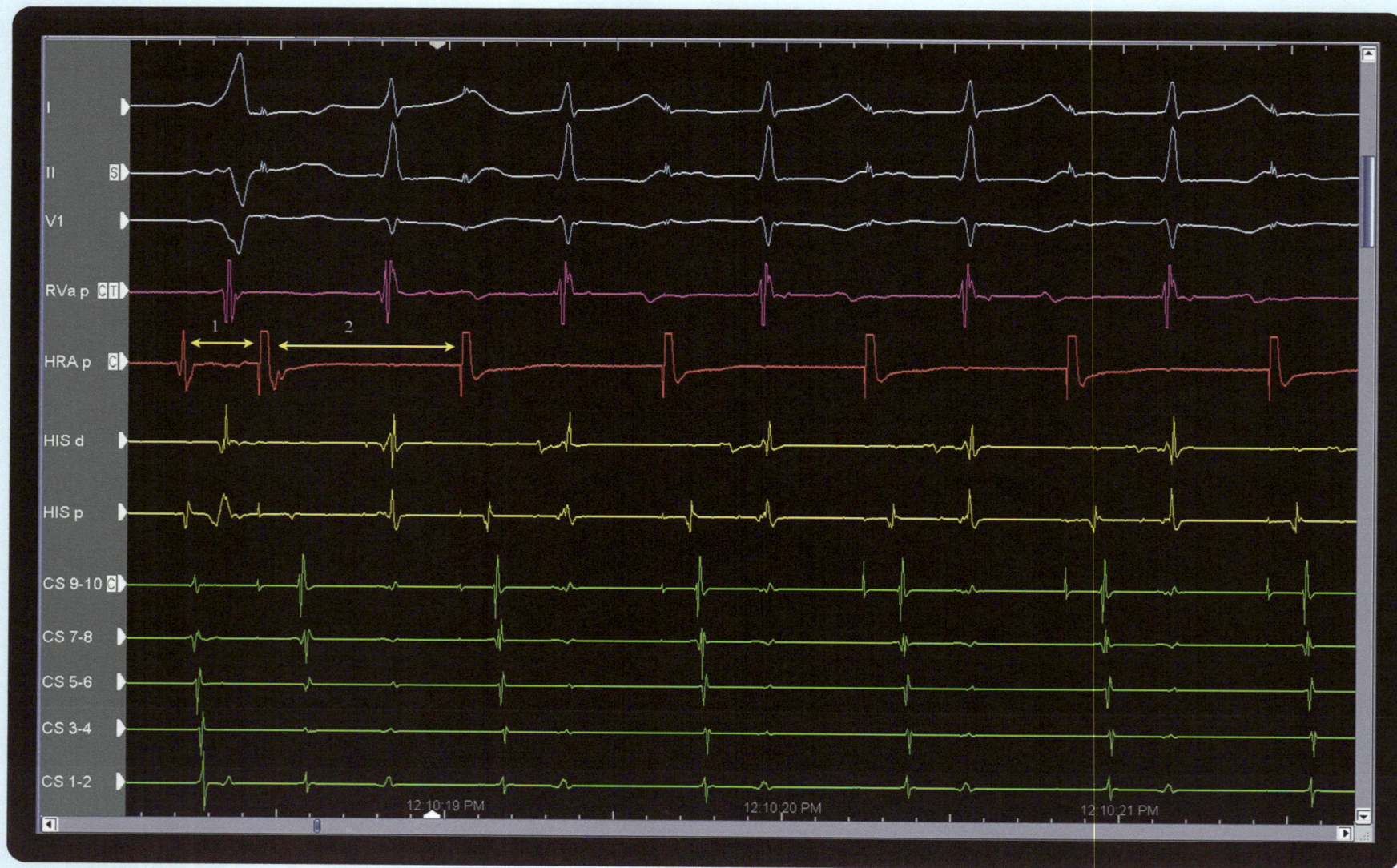

Discussion

The preexcited first cycle is during sinus rhythm. The first paced beat was tightly coupled to the preceding sinus beat and the interval between the preceding atrial depolarization and the one related to the atrial extra-stimulus was less than the ERP of the accessory. Conduction therefore proceeds exclusively over the AV node generating a narrow-complex QRS. The A-H interval of these normal beats is quite long and thus consistent with an AV nodal "slow" pathway.

The remaining beats remain narrow most likely because the pacing cycle length is shorter (faster rate) than the refractory period of the accessory pathway. This could be ascertained by knowing the ERP of the AP as determined during the rest of the study.

It is also conceivable that concealed retrograde conduction into the accessory pathway occurs after the narrow cycles, which doesn't conduct completely over the AP, but nonetheless still renders it refractory to the next occurring paced cycle (i.e., "concealed retrograde conduction" into the AP). This explanation would be more tenable, if the accessory pathway had a shorter refractory period (as assessed elsewhere during the study) than the pacing cycle length observed here that conducted without preexcitation.

This patient nonetheless has a relatively long ERP of the accessory pathway and would be very unlikely to have sudden death related to this pathway.

3.10 Unusual Initiation

Reentrant tachycardias require conduction delay in the antegrade limb of the circuit, which allows sufficient time for recovery of refractoriness in the retrograde limb or the "return cycle" of tachycardia. This tracing shows the onset of AVRT utilizing a left lateral AP. Where is the antegrade delay?

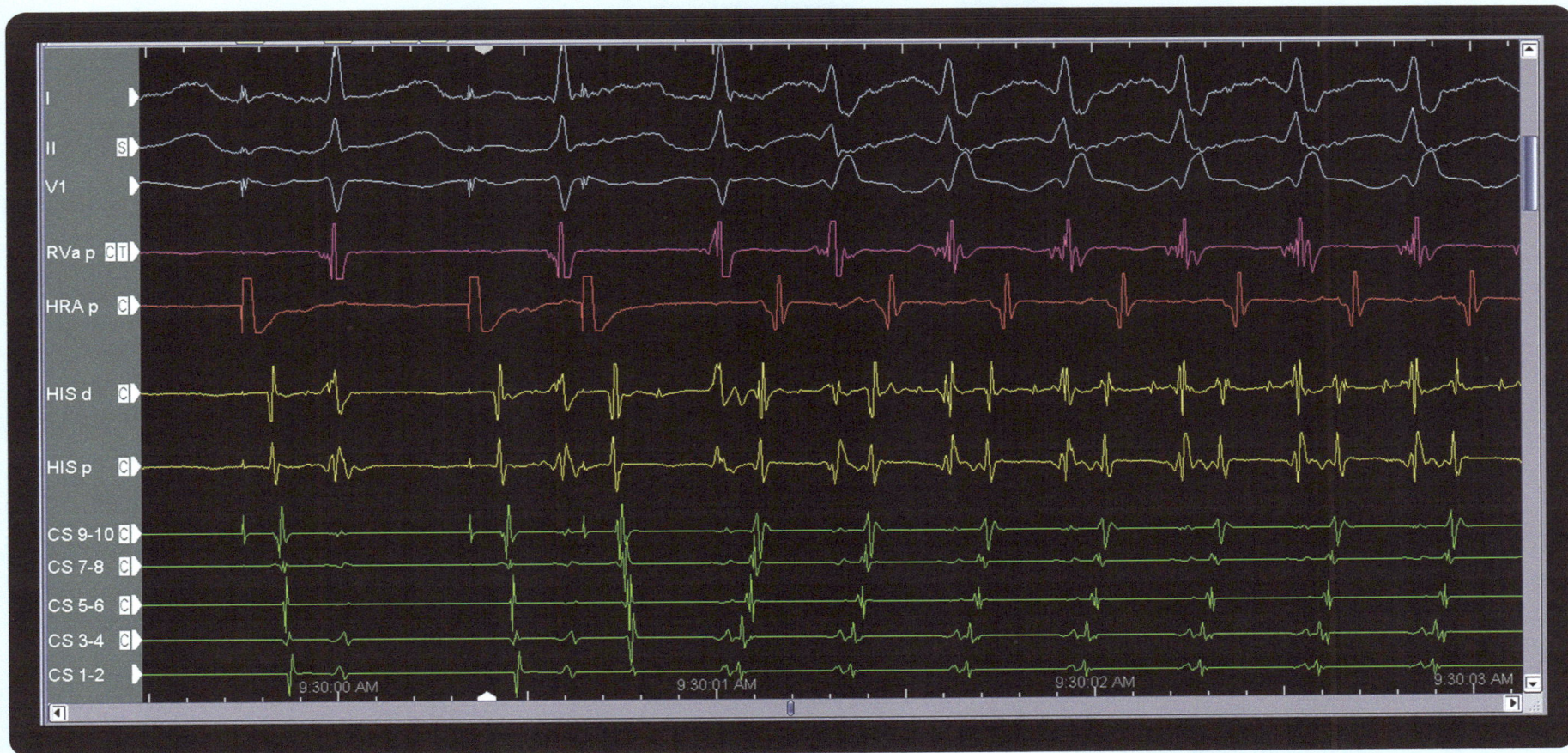

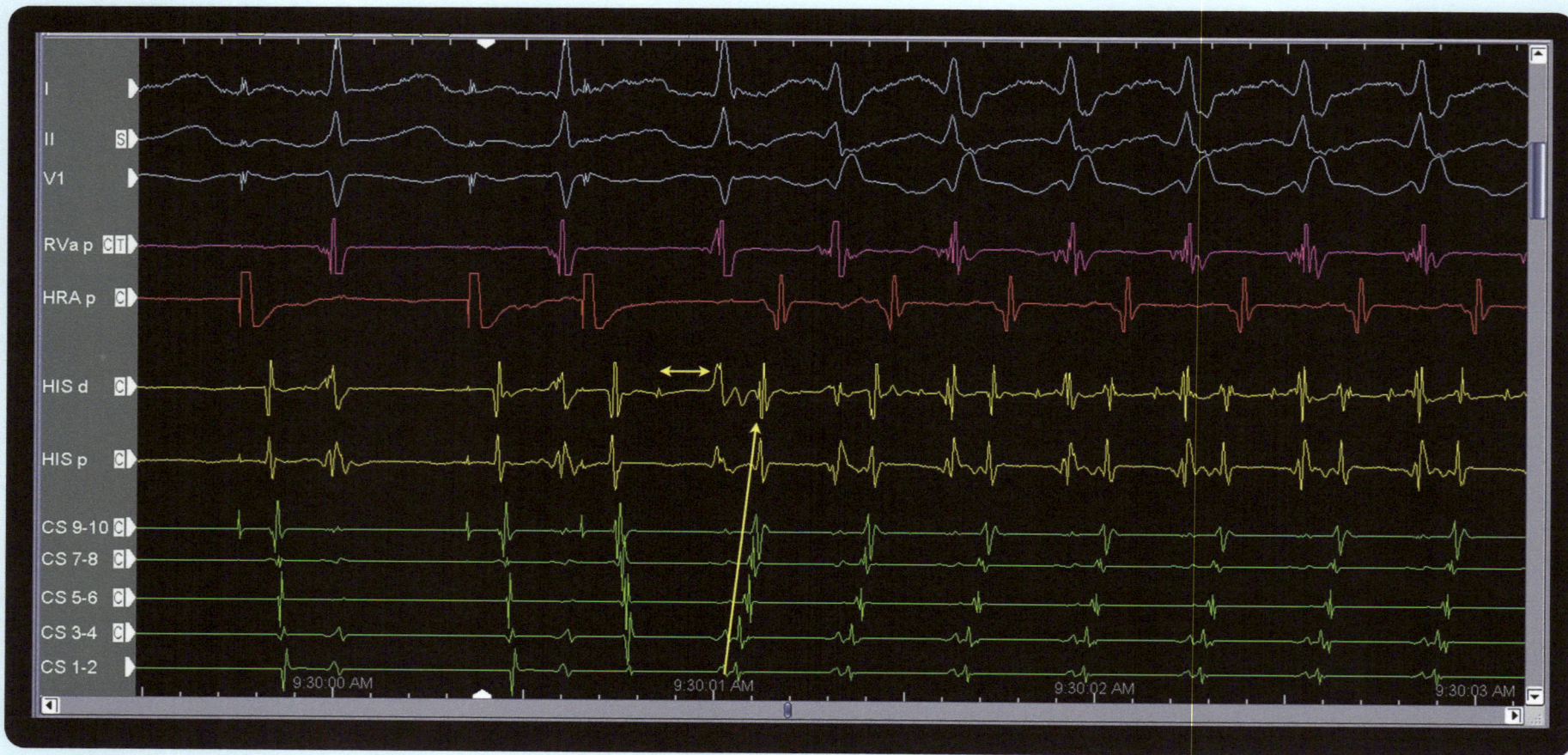

Discussion

During induction of SVT by atrial extra-stimulus pacing, as in this example, antegrade delay is *usually* related to an A-H prolongation because of the "decremental" properties of the AV node. In this example, the bulk of the delay is related to an H-V prolongation.

Usually, H-V prolongation or block is not generally seen because prolongation of A-H conduction prevents the attainment of sufficiently short H-H intervals to cause delay or block. In this example, minimal delay in the A-H interval allowed the attainment of a sufficiently short H-H interval to prolong the H-V interval. This is rarely seen as the initiator of AVRT.

A follow-up question is: Will the observed RBBB affect the V-A interval? The answer is "No," because the right bundle system is not involved in the tachycardia circuit. This is known as contralateral bundle branch block.

The circuit includes the left lateral AP, the atrium between the AP and the AV node, the AV node and His bundle, the left bundle system and the left ventricle.

3.11 AVRT Breaks

This tracing shows ongoing AVRT utilizing a left-sided accessory pathway. Notice that CS 5-6 has the earliest atrial activation. A routine maneuver is to introduce PVCs into the tachycardia. What effect will a PVC have?

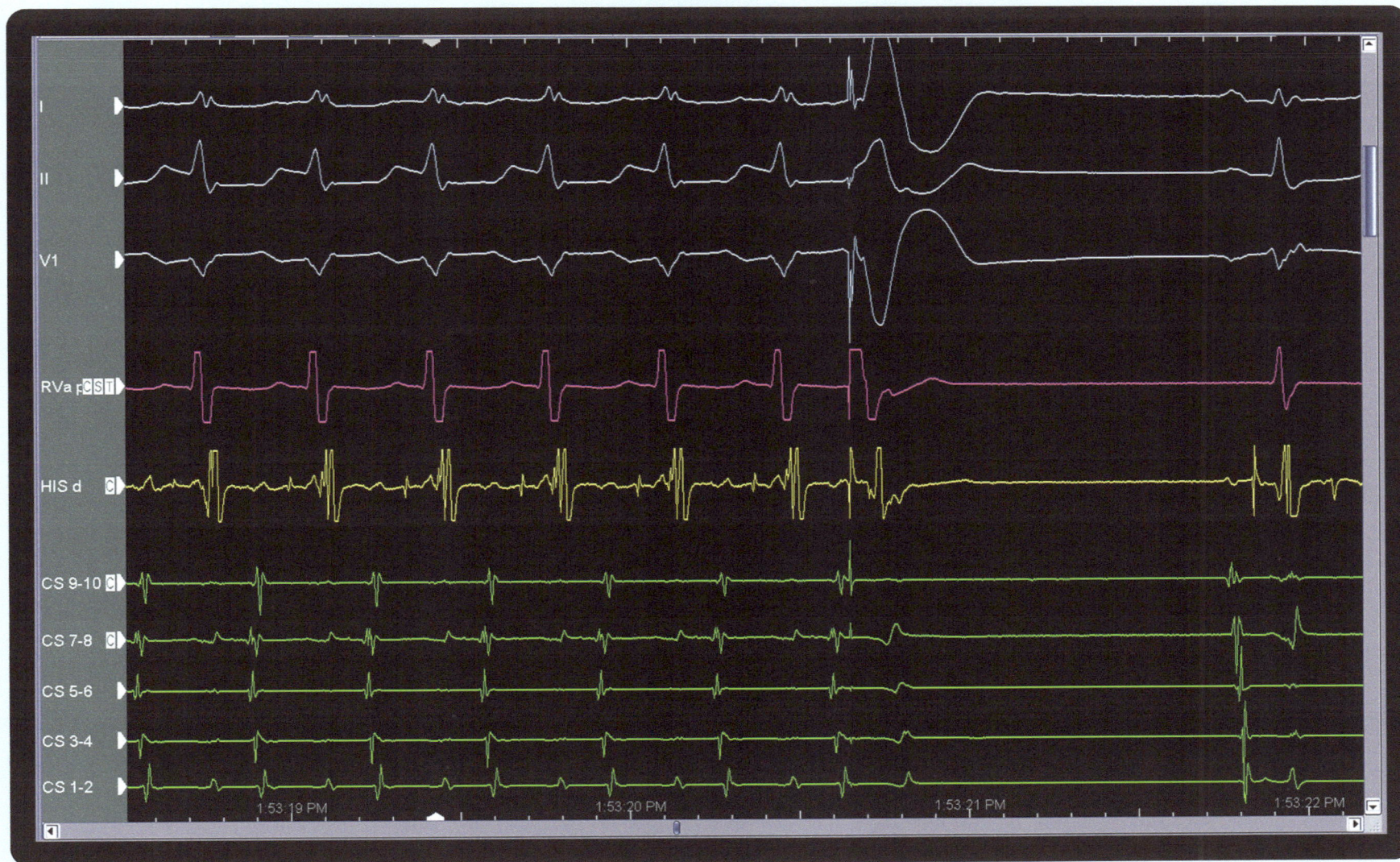

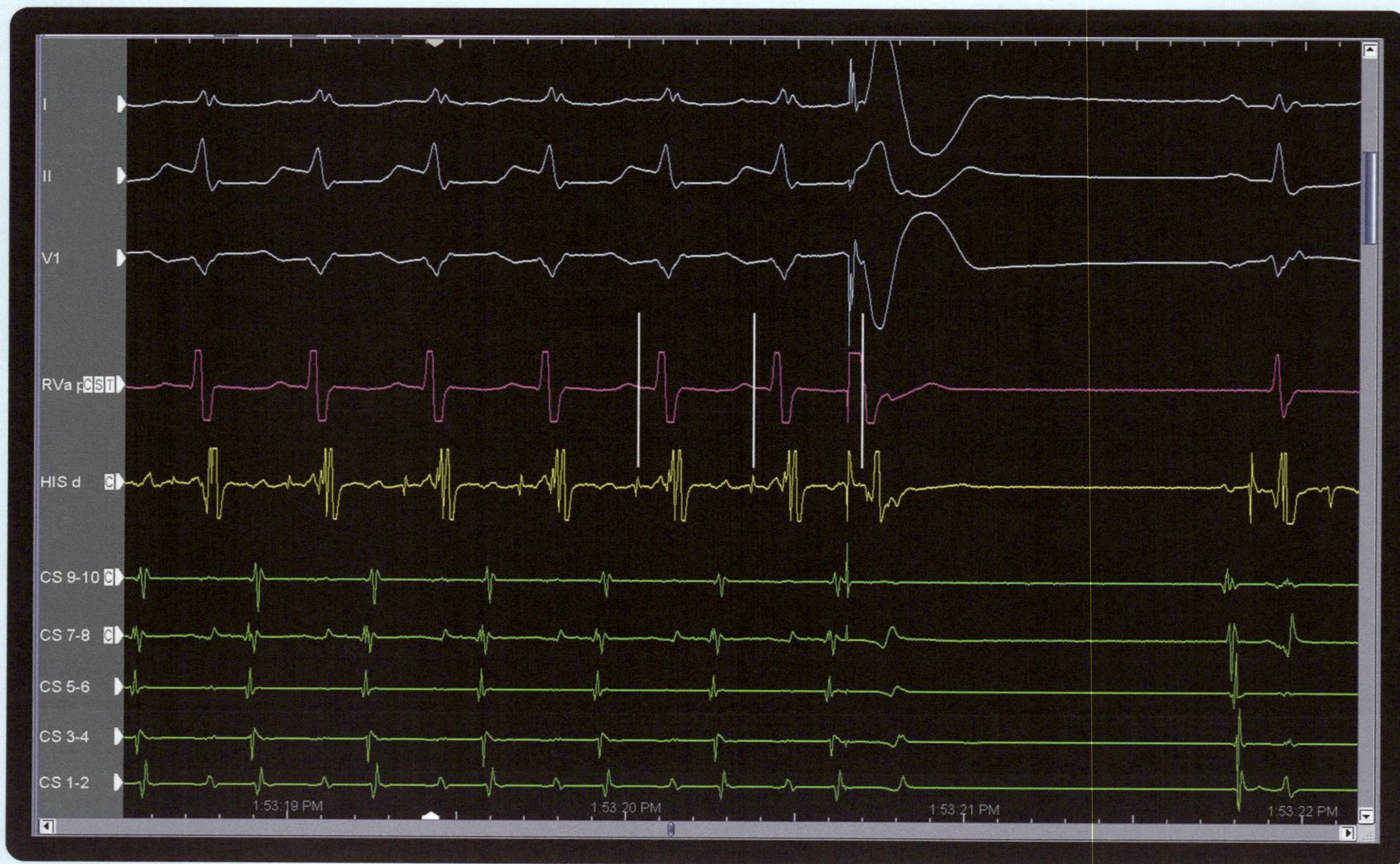

Discussion

The ventricular extra-stimulus results in a PVC that interrupts the tachycardia. It arrives at a time when the His bundle has already been depolarized (i.e., a "His refractory PVC"). This PVC breaks the tachycardia without getting to the atrium!

Breaking the tachycardia from the ventricle without conduction to the atrium rules out AT regardless of how premature the PVC is.

A His refractory PVC that advances the next "A" is proof of the existence of an AP. It does not prove that the AP is an active participant in the tachycardia circuit. However, a His refractory PVC that breaks the tachycardia without getting to the "A" is absolute proof that the accessory pathway is participating in the tachycardia circuit!

3.12 Ipsilateral BBB

This tracing shows AVRT using a left-sided accessory pathway as evidenced by CS 5-6 displaying the earliest atrial "A." Can you explain the change in the ECG morphology?

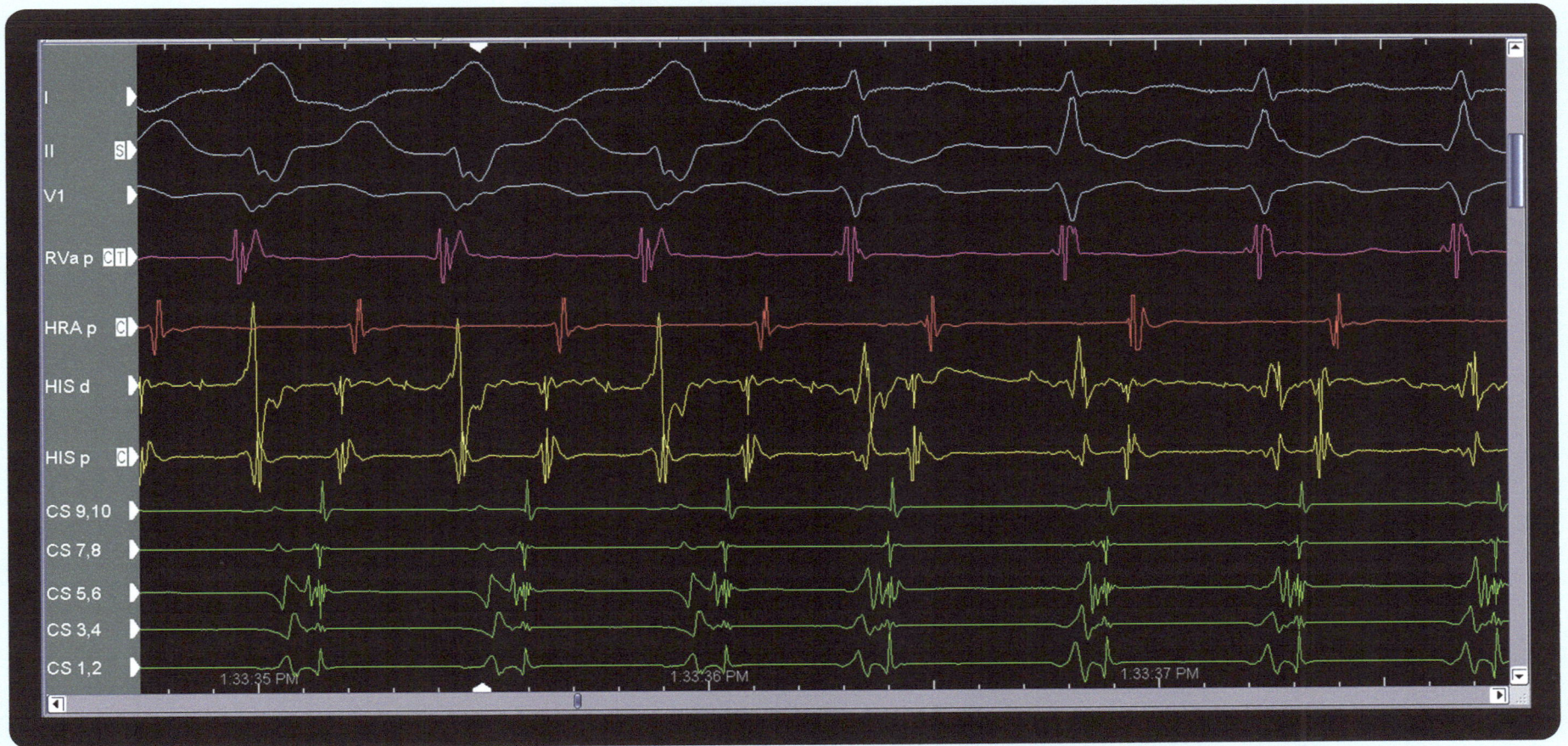

151

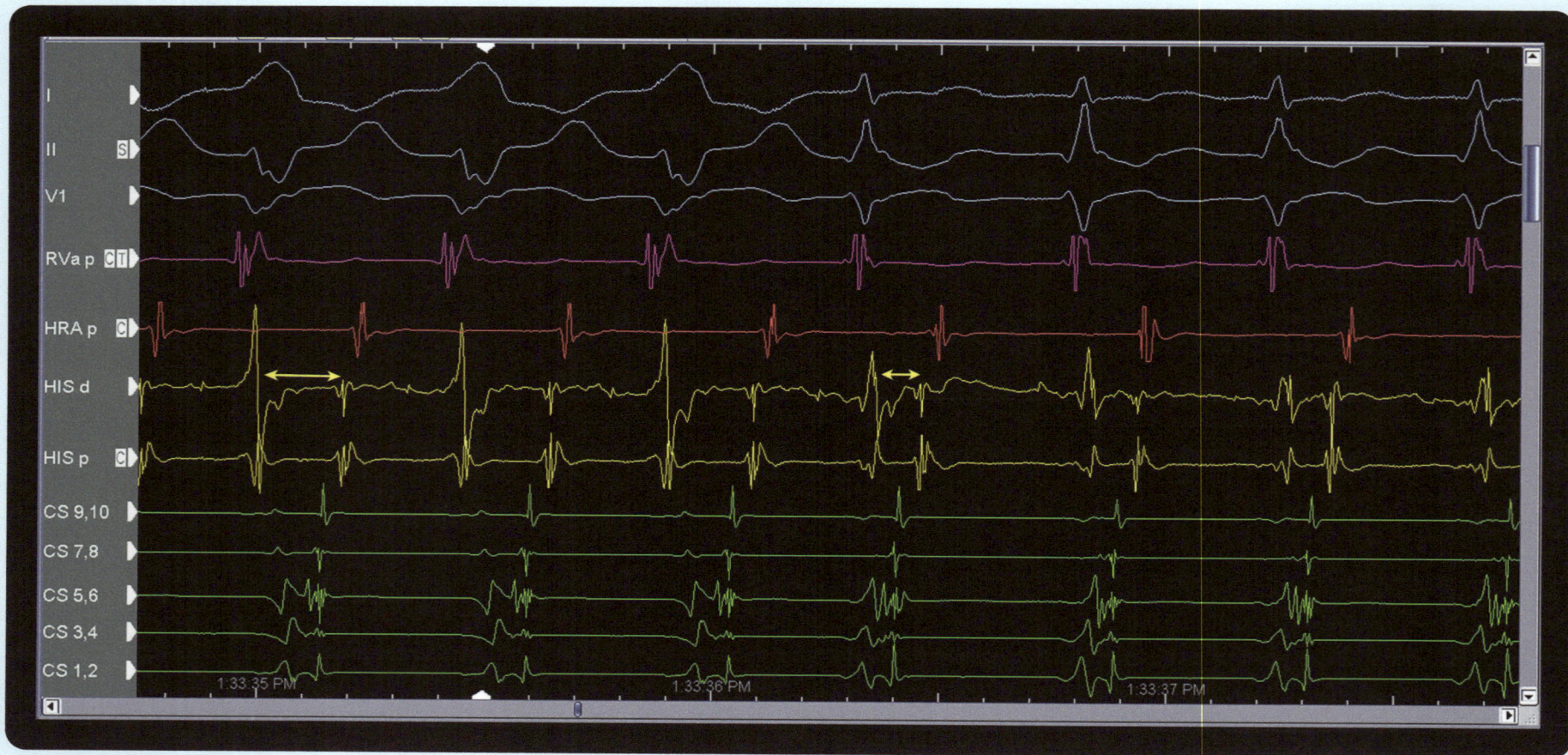

Discussion

The first 3 cycles demonstrate a LBBB pattern, which spontaneously normalizes at the fourth cycle. Compare the V-A time during the LBBB cycles and the normal QRS cycles (2 *yellow arrows*). Note that the V-A time during LBBB is longer than the V-A time during the normal QRS cycles.

This indicates that the left bundle branch *must* be involved in the circuit.

LBBB makes the circuit longer than with a normal QRS (and hence increasing V-A time). With LBBB, the circuit must return to the left-sided AP via the right bundle branch and transseptal conduction, a much longer distance than if it were getting to the pathway via the left bundle branch system. This is known as "ipsilateral bundle branch block."

We refer to the transition from a LBBB pattern to a normal QRS as spontaneous normalization.

Spontaneous normalization of the QRS most likely occurs by multiple mechanisms, but may occur because the left bundle system (as well as the right) respond to increases in heart rate and increasing sympathetic tone with shortening its ERP and enhanced conduction velocity, allowing resumption of conduction through the left bundle system.

3.13 Circuit Direction

This tracing provides several interesting observations!
Explain each beat carefully, one at a time.

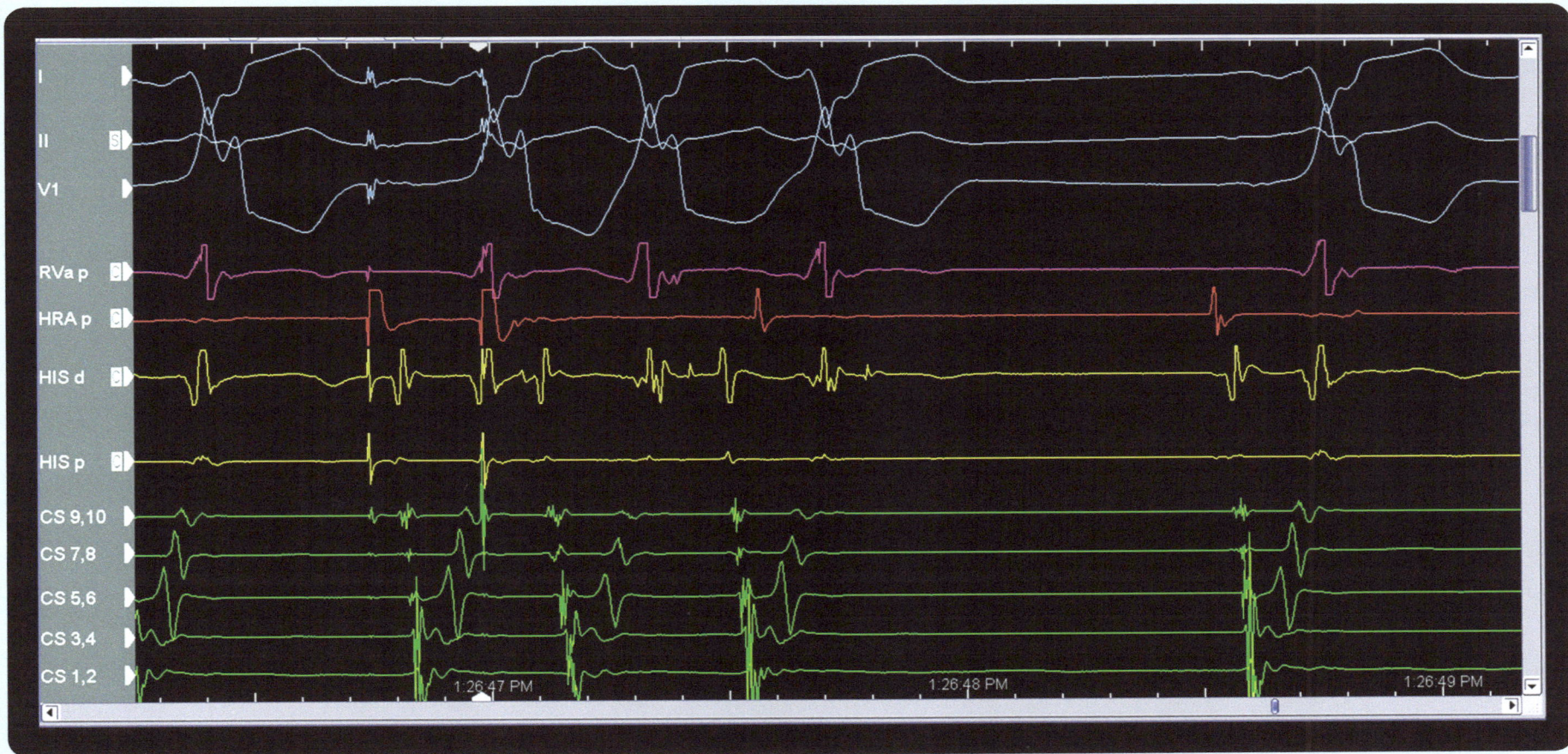

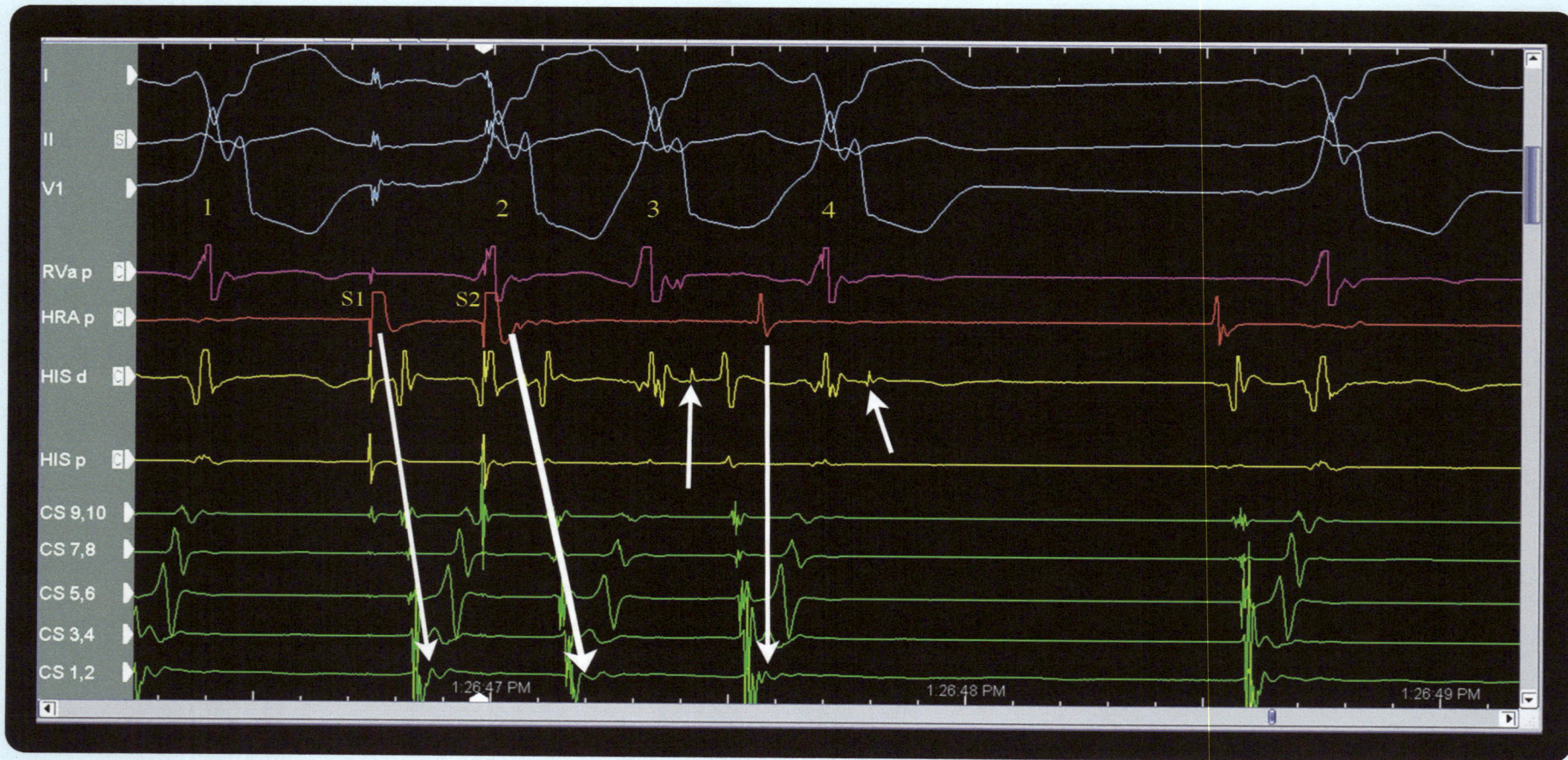

Discussion

The atrial S1 (*first arrow*) conducts to the ventricles with maximum ventricular preexcitation over a left lateral AP (the VEGM at the lateral LV, CS 1-2, is earliest and the QRS morphology is wide and compatible with exclusive accessory pathway conduction).

The S2 produces an identical QRS (*third one*) but now results in an atrial "echo" preceded by a retrograde His deflection (*first small arrow*). The retrograde "H" is probably delayed and visible because of retrograde LBBB related to prematurity, which requires the impulse to return to the normal AV conduction via the distal RBBB (a longer circuit).

The retrograde activation over the normal A-V conduction system then results in the last QRS, which again is followed by a His deflection but does *not* conduct to the atrium. Thus, retrograde block in the AV node prevents the circuit from turning around and proves the participation of the AV node in this brief "tachycardia."

3.14 Circuit Direction 2

This tracing is similar to the previous tracing. However, there is an interesting observation in the circuit. What is it?

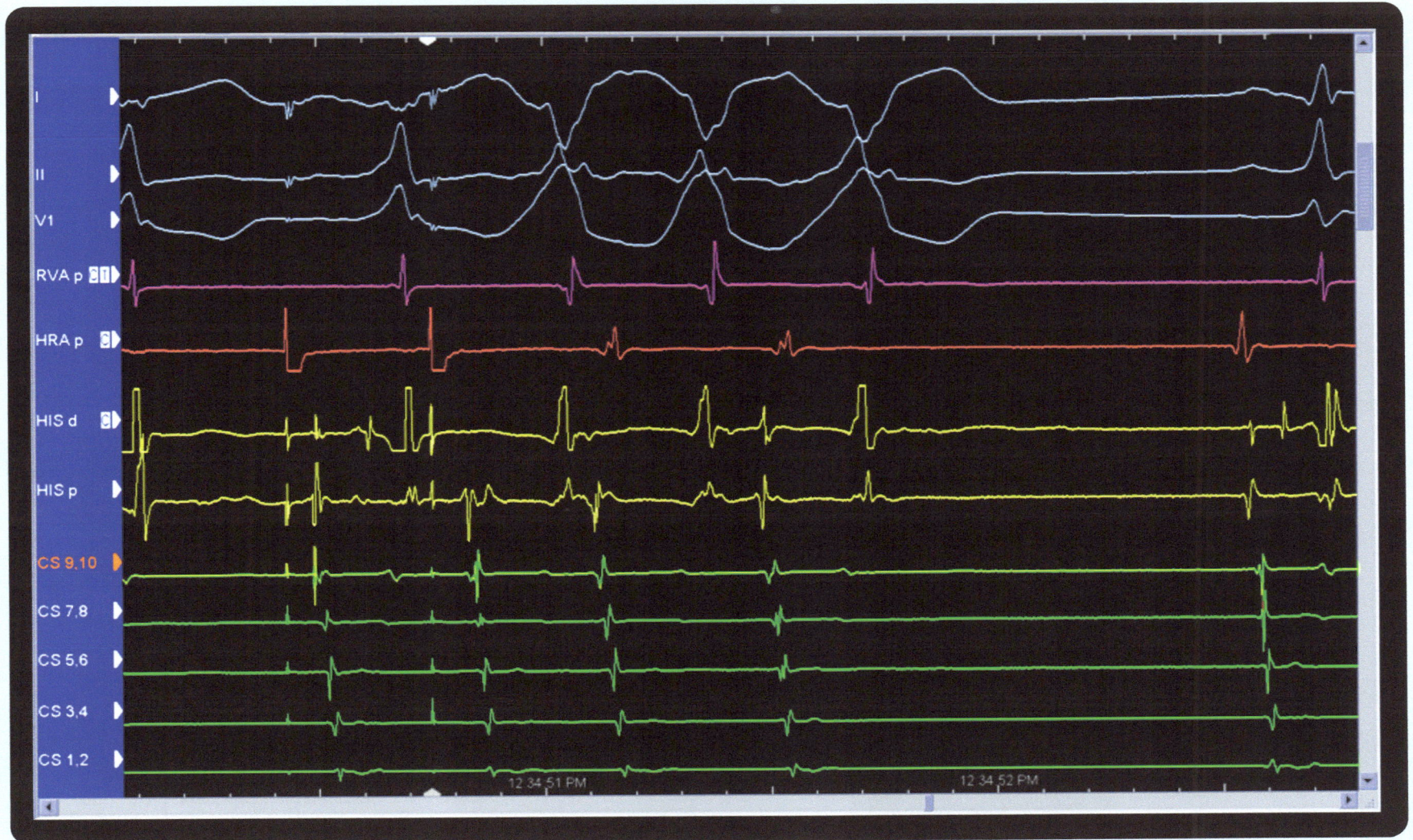

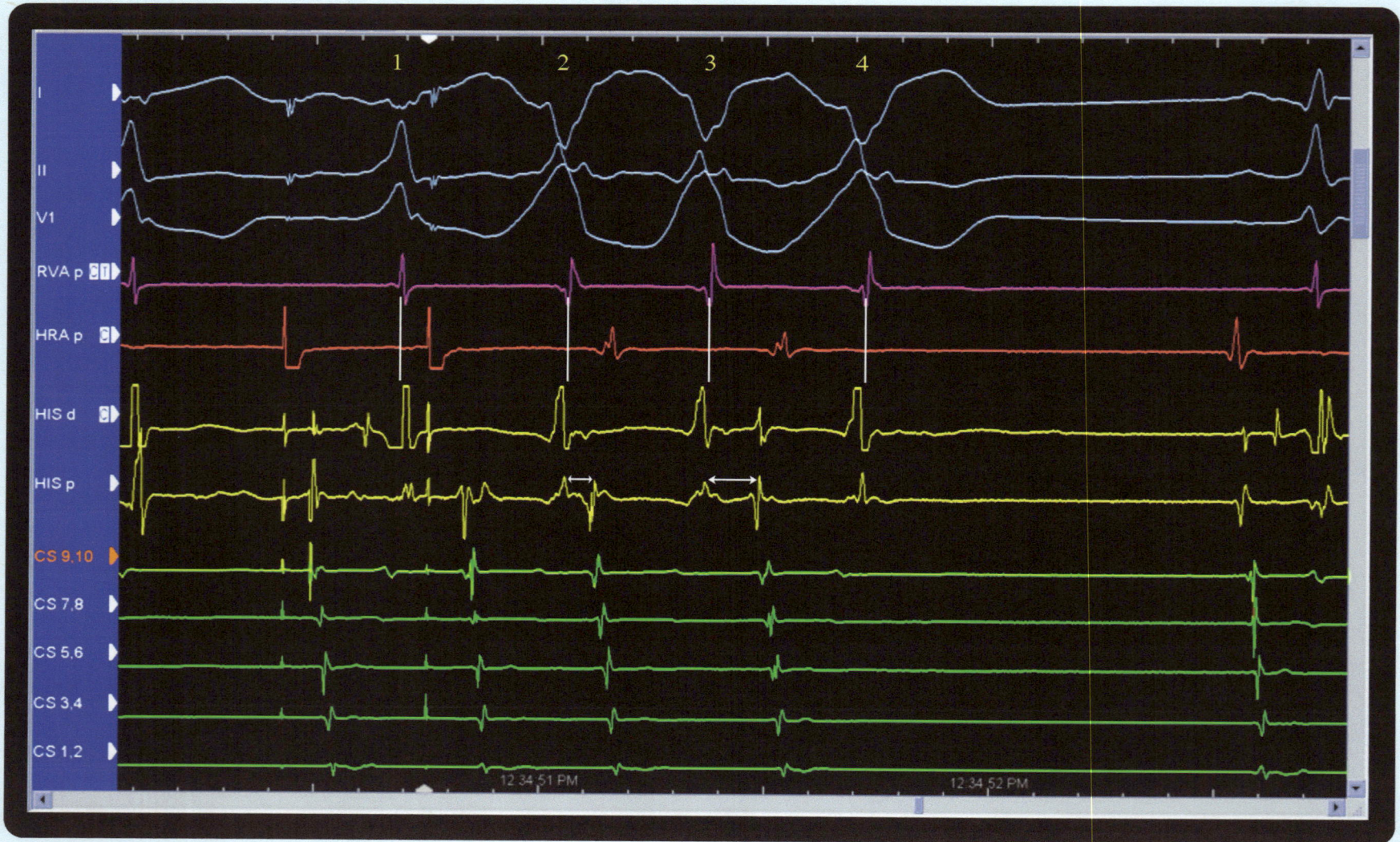

Discussion

The first cycle is preceded by a His bundle electrogram but is also preexcited (short H-V) and thus is a fusion of depolarization between the normal A-V conduction system and the AP.

The next 3 beats have no preceding His and are "maximally preexcited" over a left-sided AP (His "V" precedes the RVA "V" and the right-axis deviation with RBBB pattern are consistent with a left-sided AP).

QRS #2, resulting from atrial pacing, is followed by atrial activation with a central atrial activation sequence, that in this instance is an atrial echo related to retrograde conduction from the ventricle. (It could be a coincidental atrial ectopic beat but that was not the case here.) The echo is followed by the next fully preexcited QRS, which is again followed by central atrial activation. The activation sequence is indeed central but the V-A interval is prolonged from the previous one (decremental conduction). This final QRS is not followed by any atrial activation or visible "H."

The central atrial activation along with decrement is most compatible with retrograde conduction over the AV node...as was the case here.

A circuit with antegrade conduction over the AP and retrograde conduction over the AV node is called "antidromic." There are 2 antidromic echoes. The last QRS does not generate a third echo. This is essentially identical to the case in the preceding figure but shows the block in the retrograde limb of the circuit (AV node) to be associated with essentially retrograde "Wenkebach."

3.15 Unusual AVRT

This tracing extends observations from the 2 previous tracings
and is better understood if the previous tracings are understood.
The concepts are advanced.

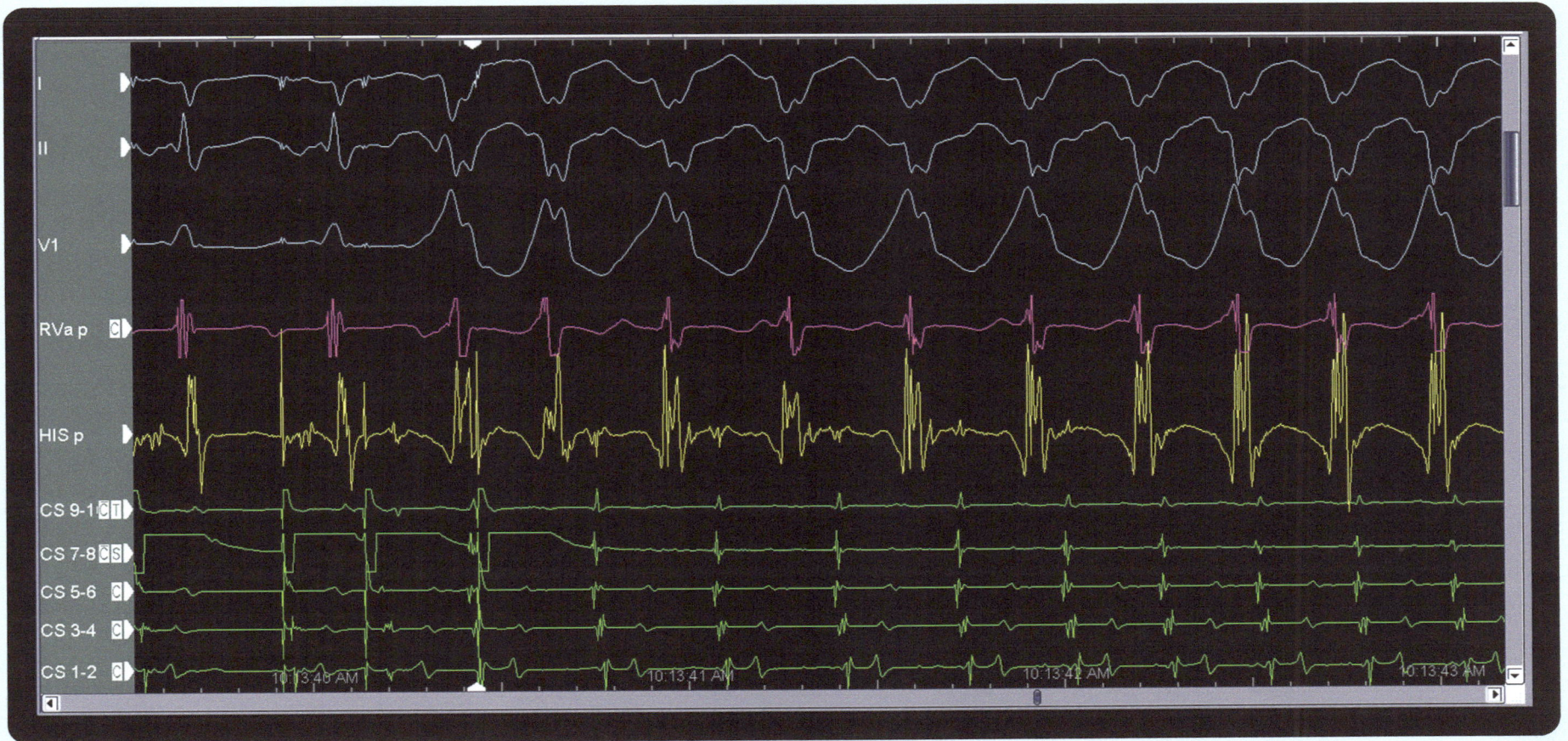

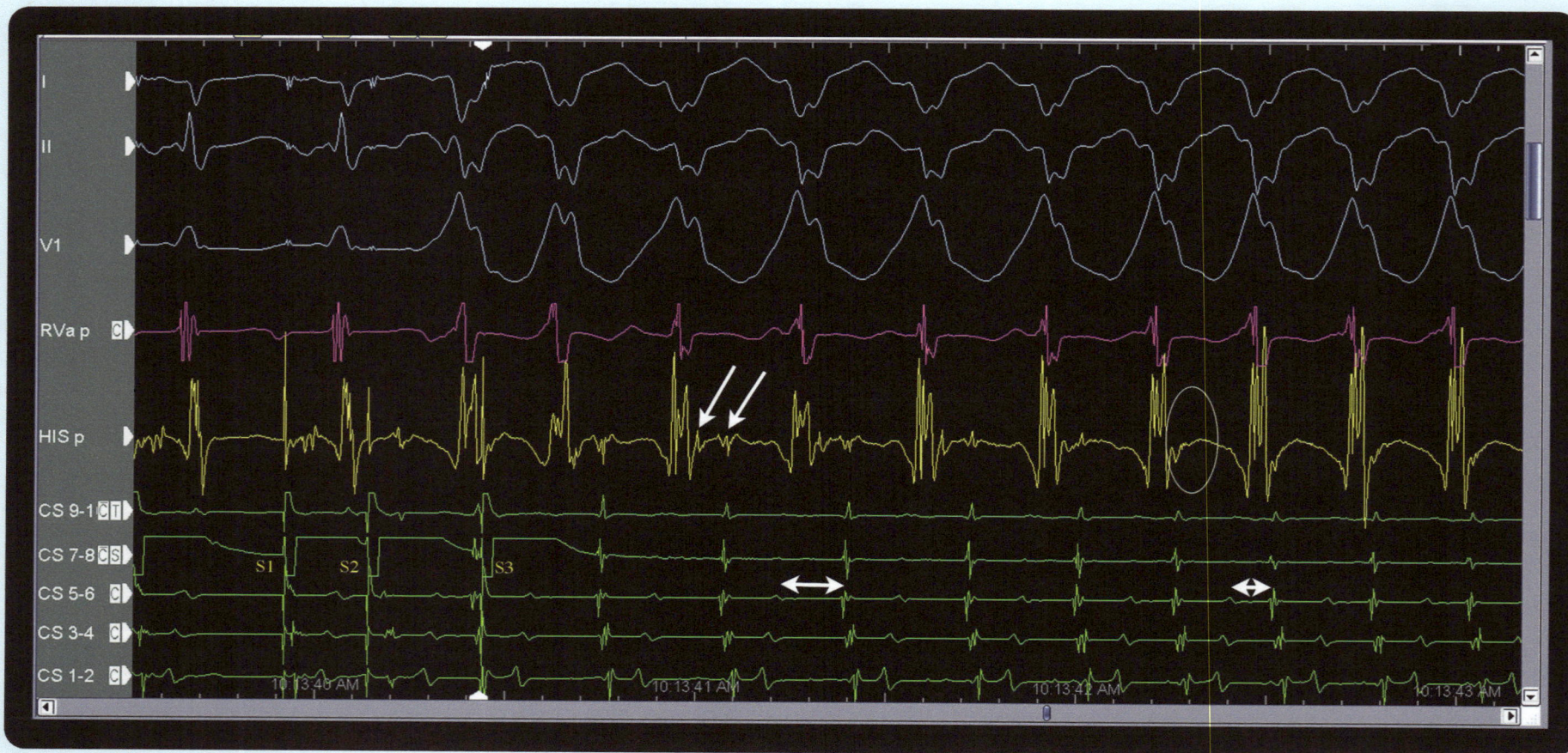

Discussion

The tachycardia in this case starts like that in the previous 2 tracings and has the same characteristics except that it keeps on going without spontaneous termination. It is a preexcited tachycardia utilizing the AP antegradely and the A-V conduction system retrogradely (antidromic tachycardia) as will be more evident from the following discussion.

The key observation is that the tachycardia cycle length (TCL) shortens midway through this tachycardia and is accompanied by, and explained by, shortening of the V-A interval (*horizontal double arrows*). The atrial activation remains concentric in both instances but a retrograde His (*first diagonal white arrow*) precedes retrograde atrial activation (*second diagonal white arrow*) during the longer cycle length tachycardia. The retrograde His is not visible with the shorter V-A tachycardia and is, in all probability, obscured within the QRS.

The longer V-A involves a longer V-H interval and is probably related to retrograde LBBB at the onset of tachycardia which causes a longer circuit back to the AV node from the AP (i.e., AP to RBB to AV node to atrium).

With spontaneous resolution of the retrograde LBBB, the circuit gets shorter (i.e., AP to LBB to AV node to atrium). The change in tachycardia cycle length related to change in the retrograde "H" to "A" interval proves that the AV node is part of the circuit (the retrograde limb of the circuit).

3.16 Unusual Break

This is an example of spontaneous normalization of the QRS and its effect on orthodromic AVRT. Why did the tachycardia break?

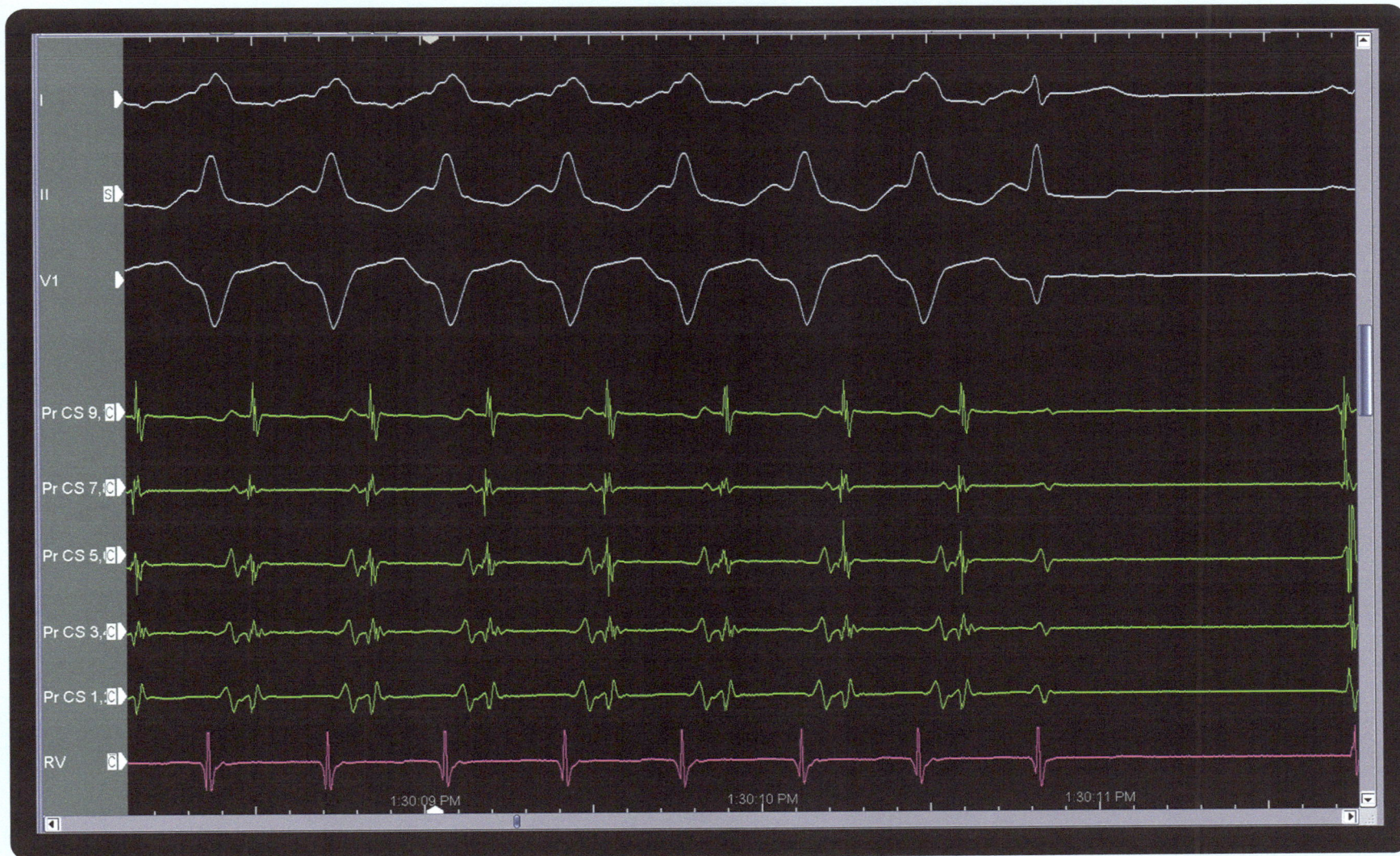

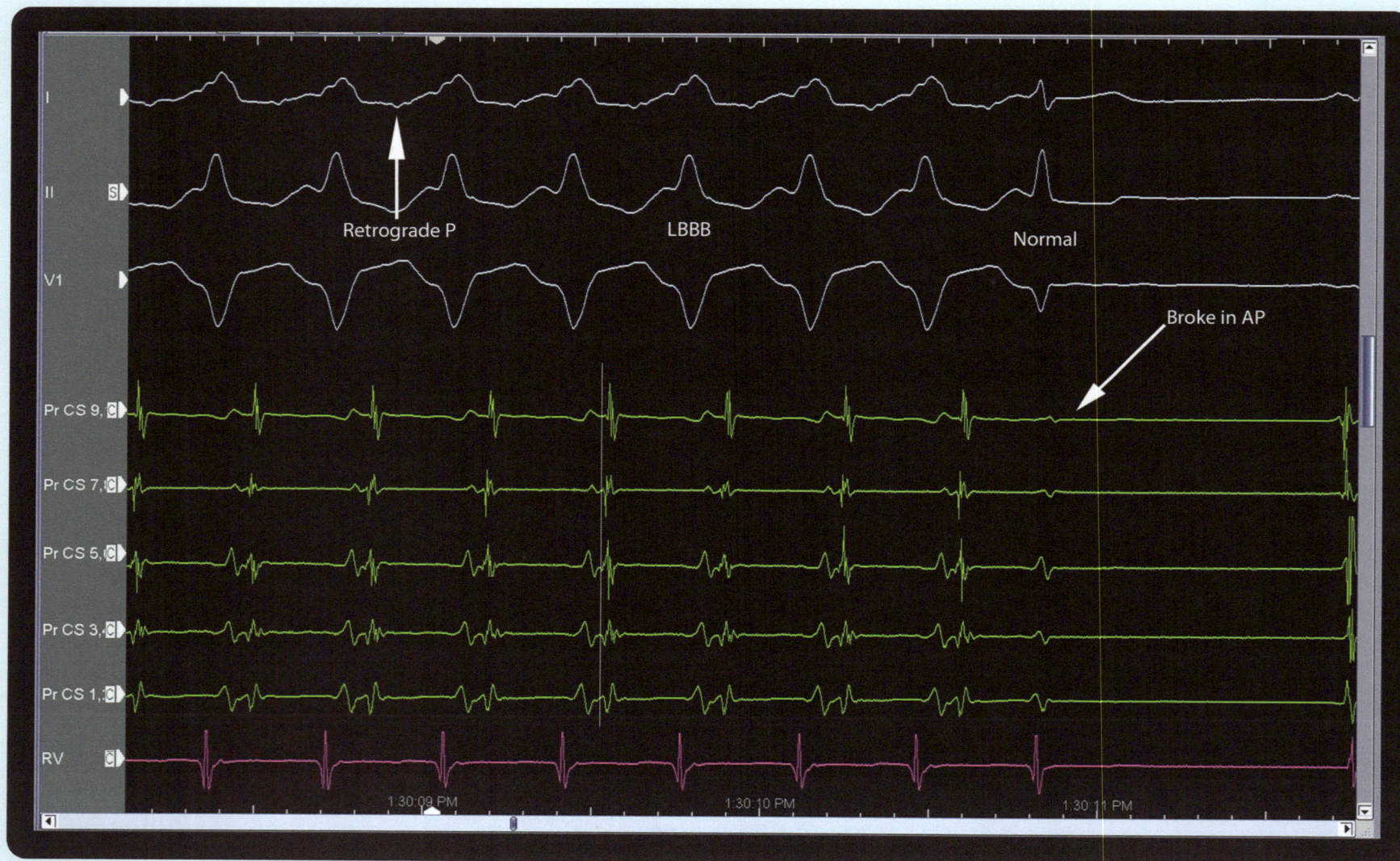

Discussion

The tracing begins with ongoing orthodromic AVRT with LBBB aberrancy over a left-sided AP. Atrial activation is eccentric and earliest at CS 1-2. The surface ECG demonstrates a LBBB pattern with normalization of the QRS complex on the final beat coincident with termination of tachycardia. This suggests that maintenance of LBBB (ipsilateral BBB) is important to the tachycardia circuit.

In the presence of LBBB, the wave front of activation during this tachycardia travels antegradely down the AV node, down the right bundle, through the septum to the LV and retrogradely up the AP to the atrium.

When LBBB recovers (refer to Chapter 1, Section 5), the tachycardia circuit becomes smaller. The ventricular component is now from the distal left bundle to the left AP. The resulting abrupt shortening of the time of arrival of activation to the AP encroaches on the refractory period of the AP, which now fails to conduct to the atrium and tachycardia terminates.

[See movie: https://bit.ly/purvesadvanced.]

Notes

Cardiac Electrophysiology 2

CHAPTER 4   AF

4.1 Differential Pacing

The most common end point for a pulmonary vein (PV) isolation procedure is the demonstration of entrance block into and exit block from the pulmonary vein.

When checking for entrance block, it may be difficult to distinguish between "far-field" left atrial (LA) potentials and "near-field" PV potentials.

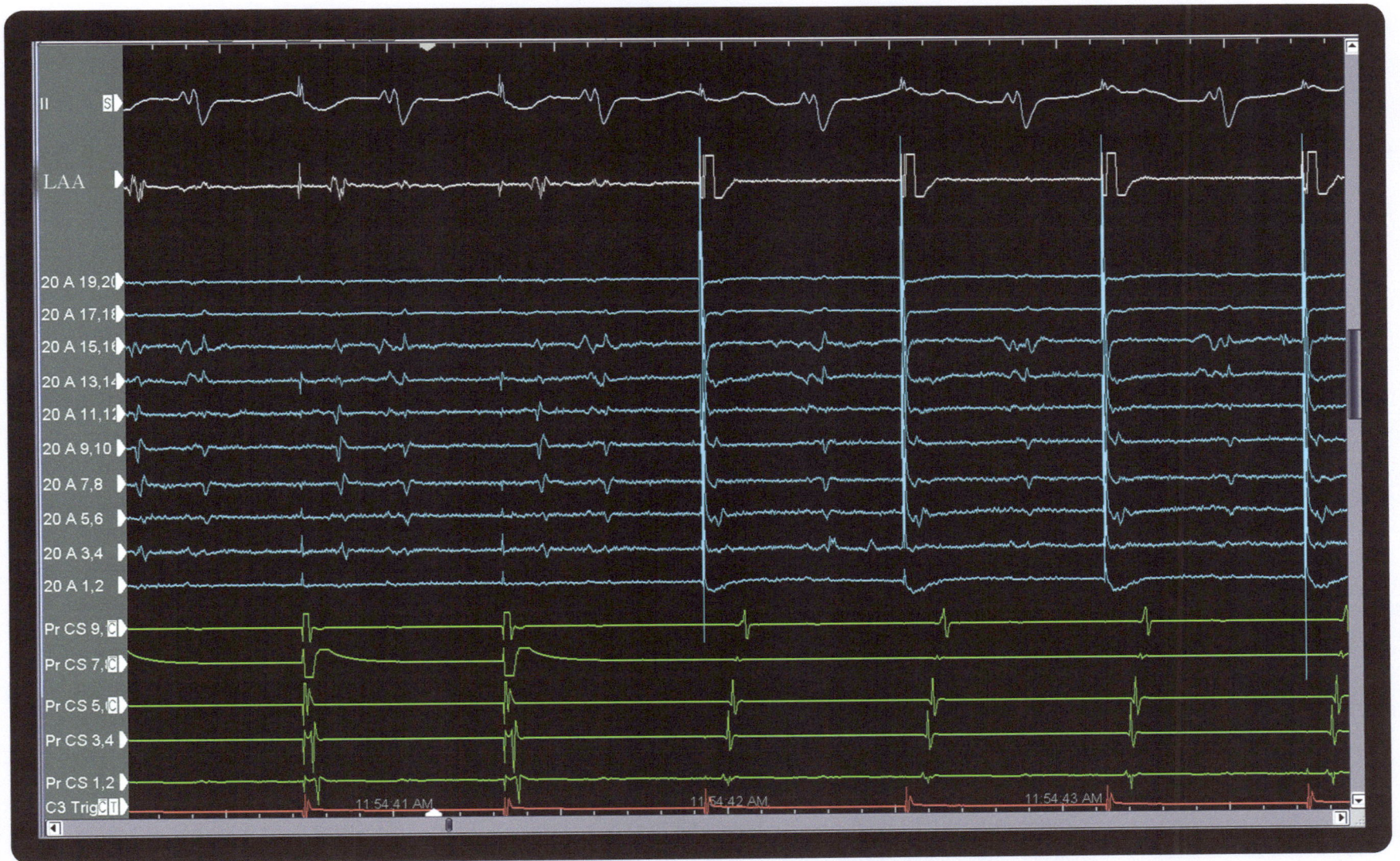

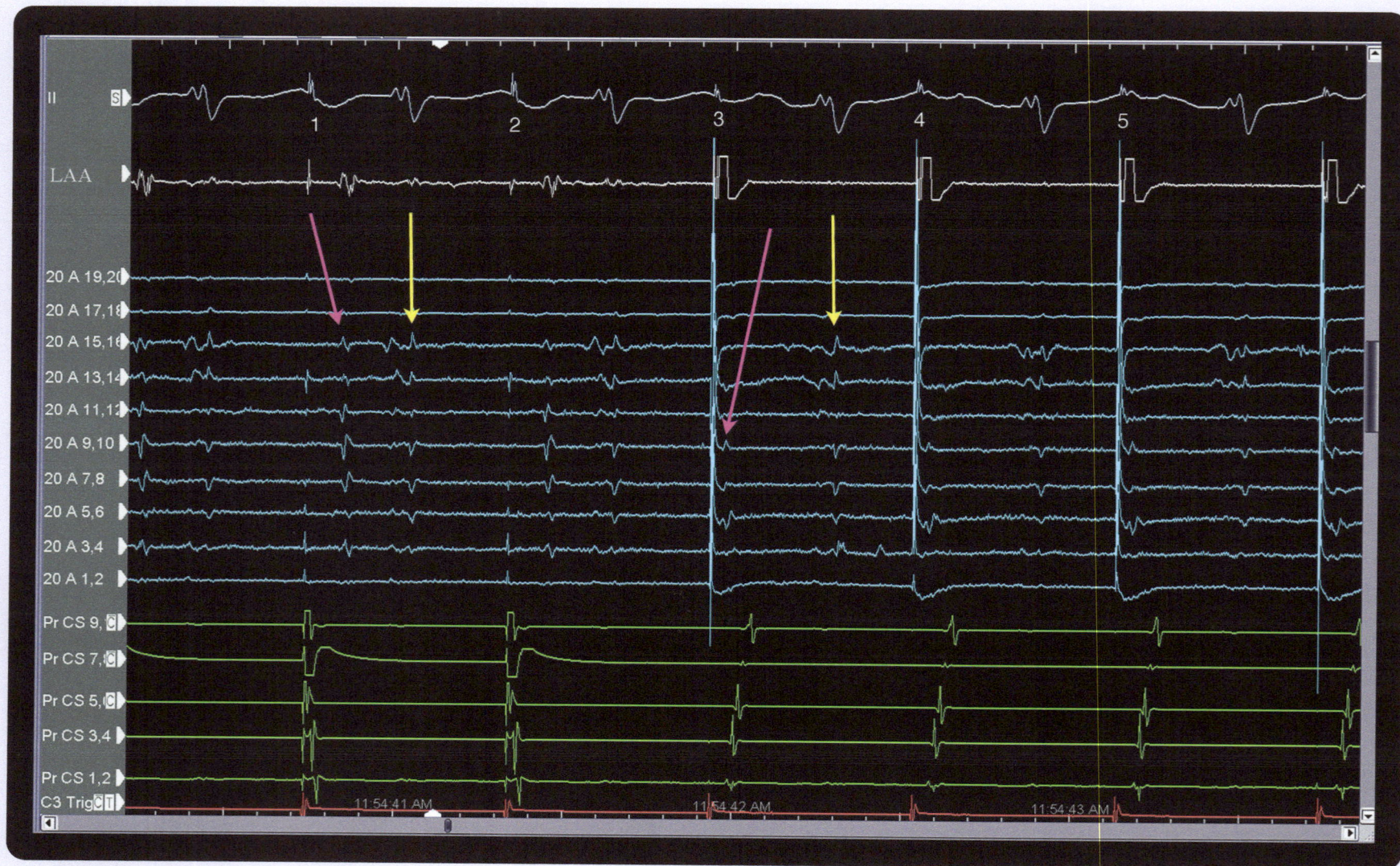

Discussion

In this example, the *magenta arrow* points to the electrograms in question (LA vs. PV) on the Lasso catheter. The *yellow arrow* points to far-field ventricular potentials.

Beats 1 and 2 are recorded during CS pacing. Beats 3 to 5 are recorded during pacing from the ablation catheter placed in the left atrial appendage (LAA). Notice that during LAA pacing, the electrograms identified with the *magenta arrow*, have been pulled in toward the pacing stimulus. This demonstrates that the source of these electrograms is "far-field" from the pacing site and not the pulmonary veins. If the electrograms had been "near-field" PV potentials, upon switching to LAA pacing, they would have stayed "late" or would not have been "pulled in."

This principle can test the source of electrograms recorded from any of the pulmonary veins.

4.2 Proof of?

This tracing shows 3 beats of sinus rhythm with 2 near-field
PV activations (*yellow arrows*) recorded from
the circular mapping catheter within a PV. What does
this observation tell us?

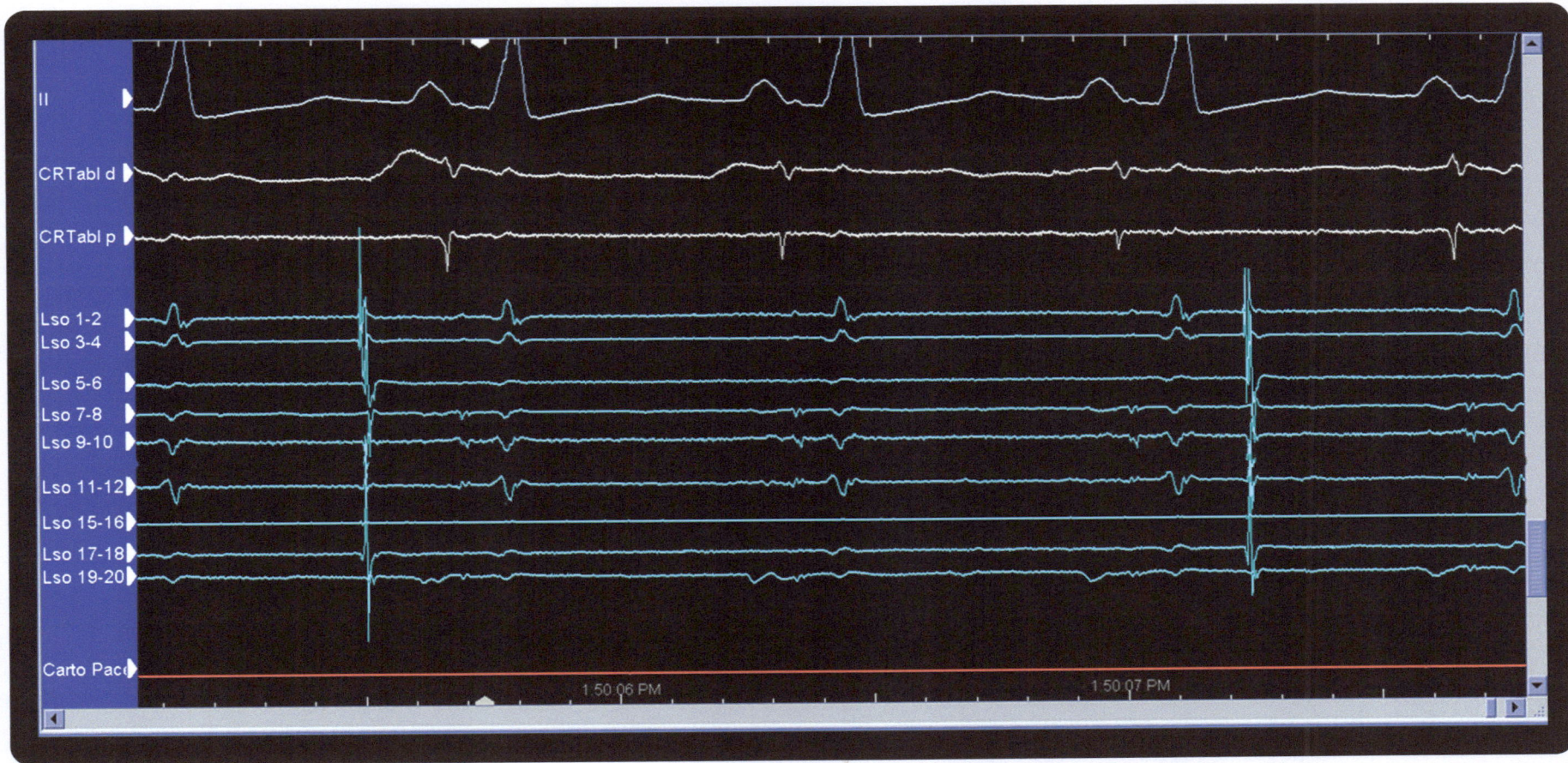

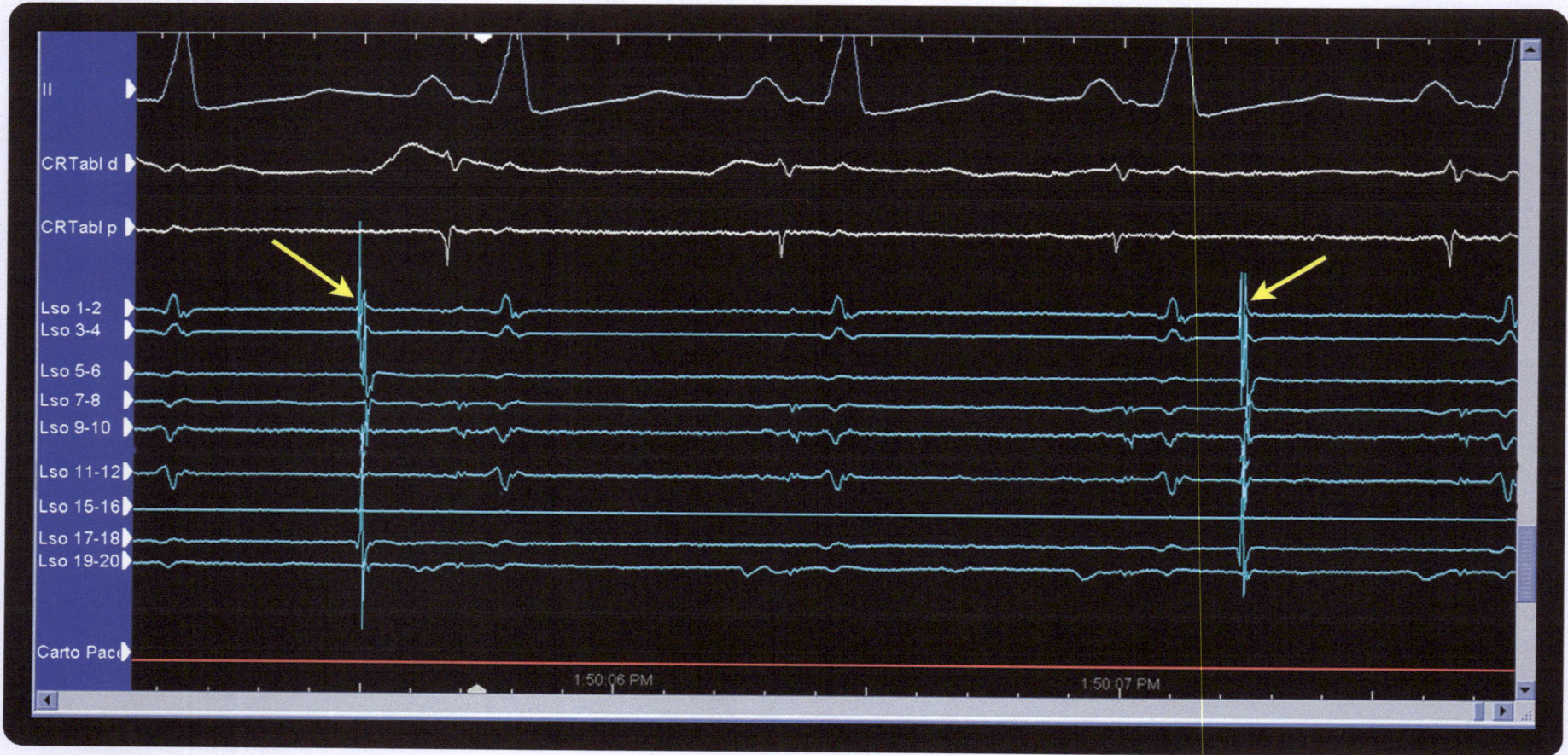

Discussion

Note that these activations do not perturb the ongoing sinus rhythm. In essence, they are independent of the surface ECG P waves, independent of the ablation catheter electrograms and independent of the far-field electrograms recorded by the circular mapping catheter (far-field "A"s and "V"s).

Isolated pulmonary veins often exhibit slow, regular spontaneous activation. When the vein activity is indeed slow, regular, and unperturbed by sinus rhythm, it suggests entrance block. Likewise, over long periods, failure to conduct out of the vein to the left atrium (LA) (resulting in a PAC) suggests exit block.

To best demonstrate exit block, the largest of the independent activity electrograms can be paced. Often convincing PV capture can be seen and the inability to drive the atrium demonstrates exit block.

4.3 Re-do

During the healing phase following PV isolation, some of the PV encircling lesions can recover resulting in gaps in the ablation circle and thus electrical reconnection of the pulmonary veins. Commonly, this occurs at 1 to 3 sites per vein or pair of veins. A repeat procedure attempts to map and ablate the gaps in the previous circle rather than redoing the entire circle.

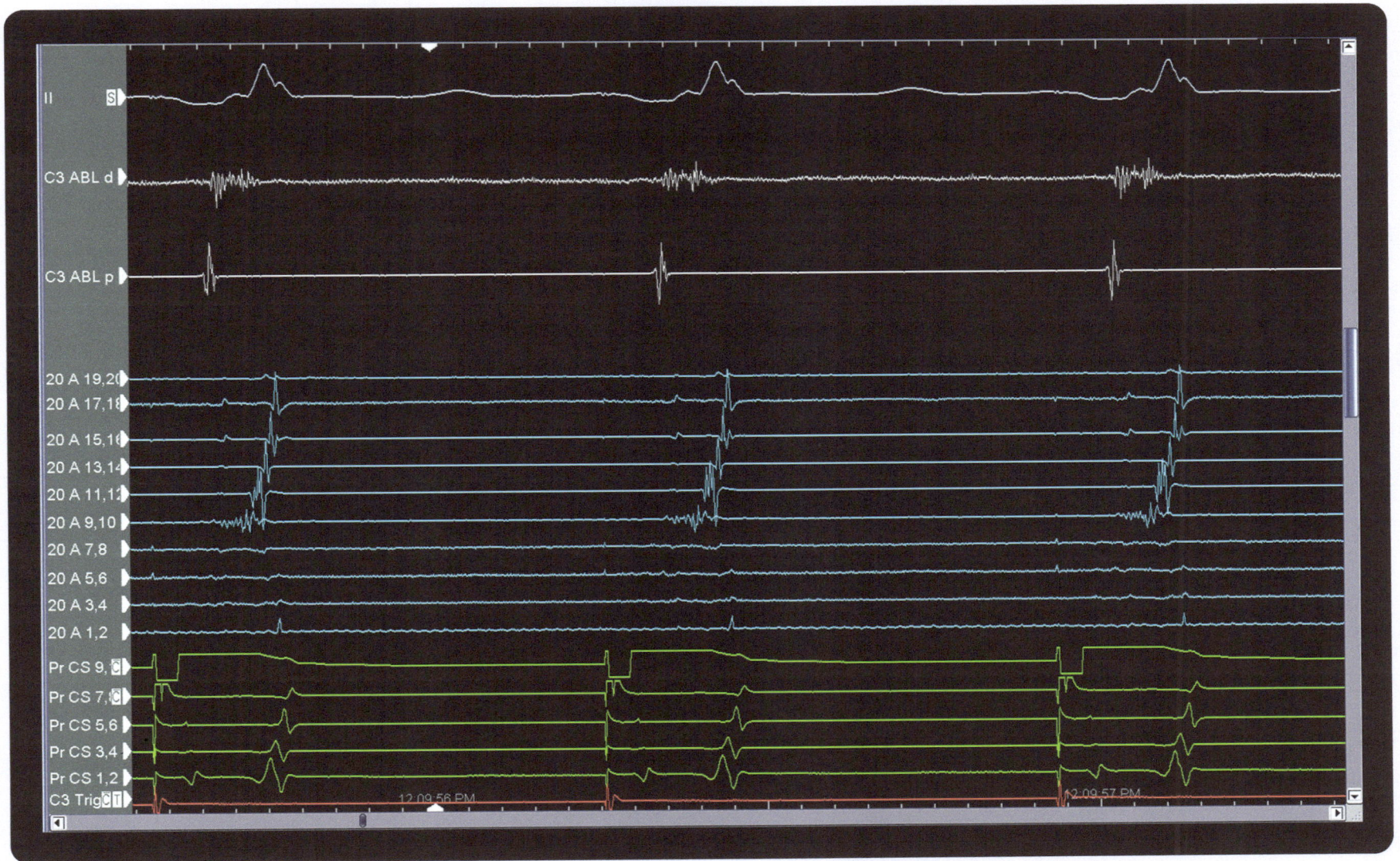

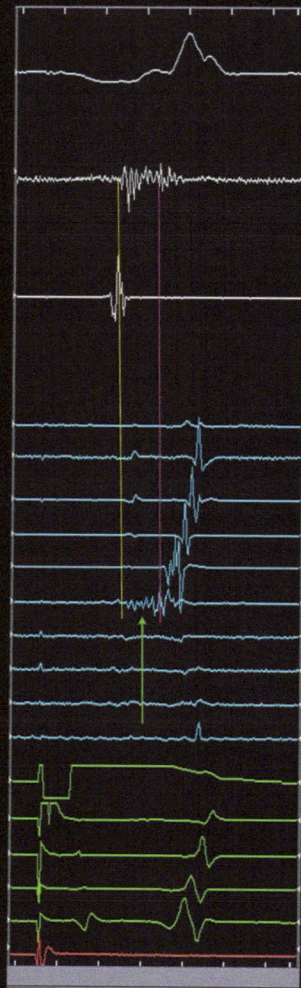

Discussion

In this tracing, the circular mapping catheter (*blue signals*) shows near-field electrograms in A9-10 to A19-20 strongly suggesting reconnection of the PV. Activation of the near-field signals shows earliest activity at A9-10 (*green arrow*) suggesting these poles are nearest the gap. Also notice that the local electrogram at A9-10 is long and fractionated indicative of a zone of slow conduction (see expanded view).

The ablation catheter, near electrode A9-10, shows a similar electrogram with fractionated continuous activity from the early LA signal (proximal ablation, *yellow line*) to the near-field PV potential (A9-10, *magenta line*).

Ablation at this site quickly re-isolated the PV.

Notes

Cardiac Electrophysiology 2
CHAPTER 5   THE UNEXPECTED

5.1 Tachycardias

To terminate a tachycardia, we often overdrive pace the tachycardia from the ventricle, as in this example. Occasionally, unexpected and somewhat surprising results occur.

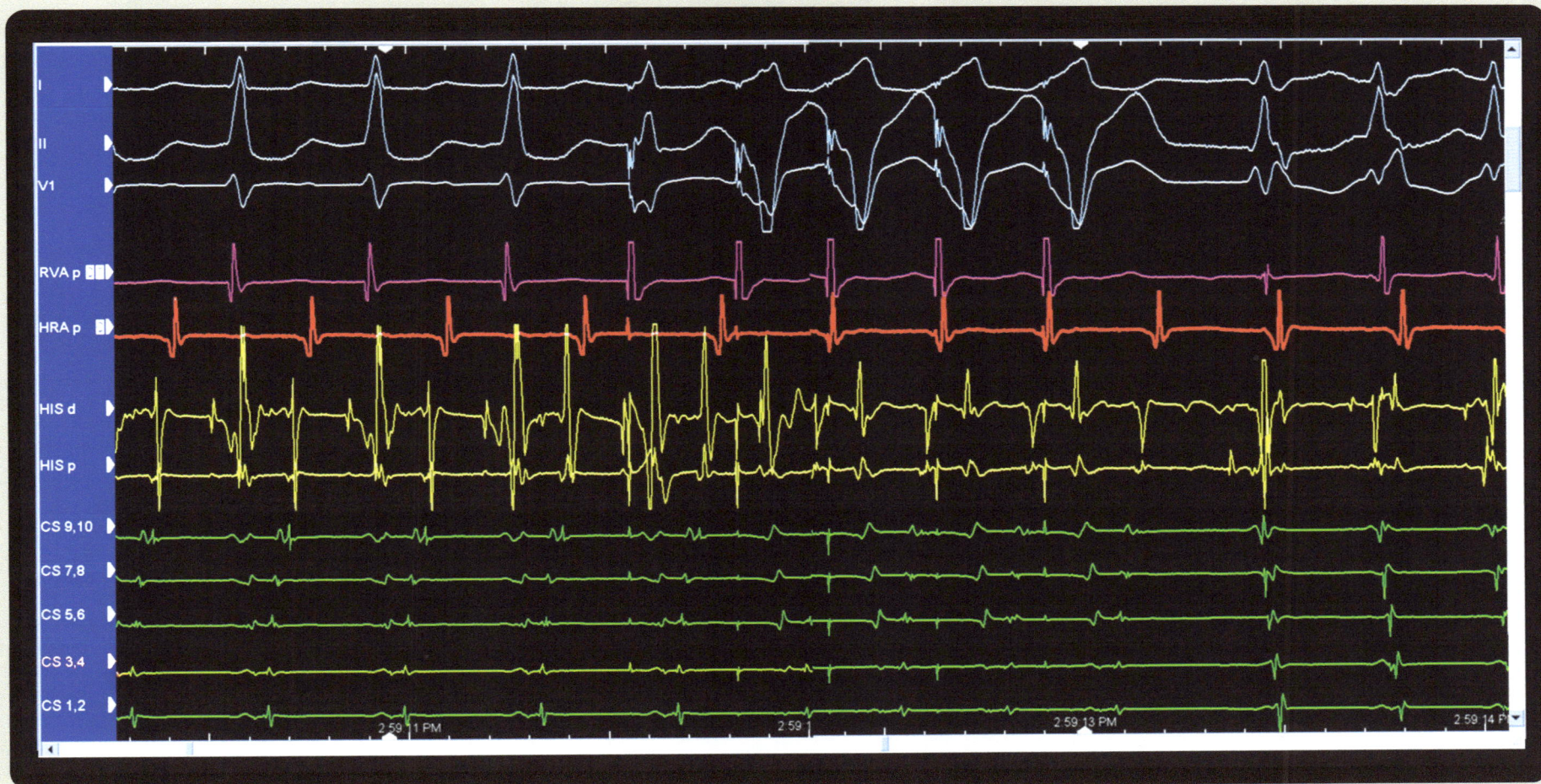

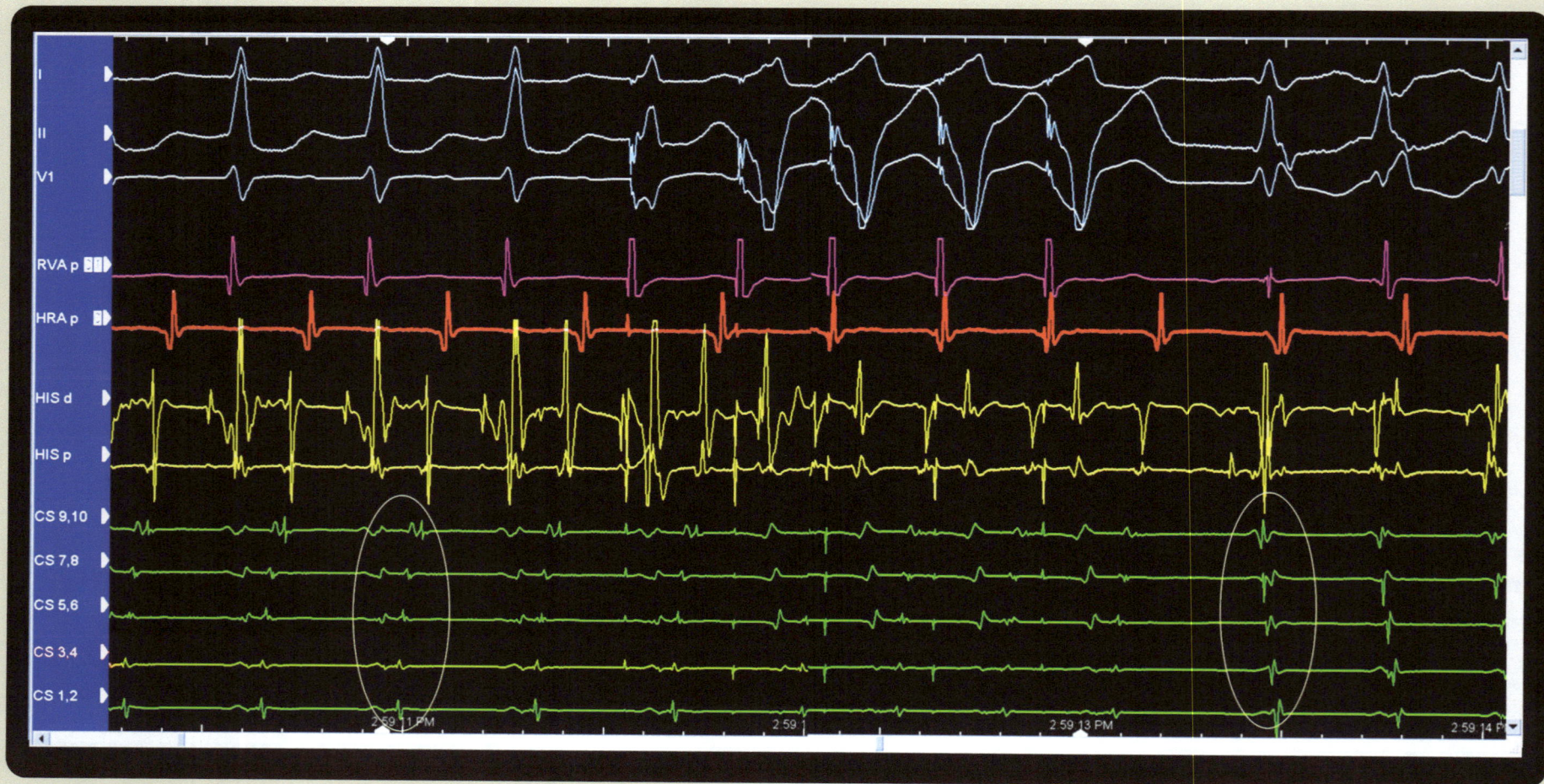

Discussion

The first 3 beats in this tracing demonstrate AVRT using a left lateral AP. Note the eccentric atrial activation pattern with earliest "A" in the distal CS 3-4. A short 5-beat ventricular burst was delivered in an attempt to terminate the tachycardia. However, sinus rhythm was not restored!

Following the ventricular burst, the tachycardia continues, *but* it has a shorter cycle length and a different atrial activation sequence. The atrial activation sequence is central (earliest "A" in the proximal coronary sinus (CS)) with a constant V-A time of zero. These findings are compatible with AVNRT!

Dual AV nodal physiology is relatively common and can coexist in patients with AVRT. In fact, some patients with AVRT have tachycardia only when the AV nodal slow pathway is used for initiation and maintenance of the AVRT. Once the AP has been ablated, it would not be unexpected to initiate AVNRT.

Note: On the surface ECG, with the initiation of AVNRT, a RBBB pattern developed and was maintained. Explain the possible mechanisms responsible for the bundle branch block, how it may be maintained and what maneuver may be able to normalize the bundle branch block. These questions were answered earlier in this book.

5.2 Tachycardias 2

During the course of an EP study, routine maneuvers occasionally lead to unexpected results. What has this single PVC done?

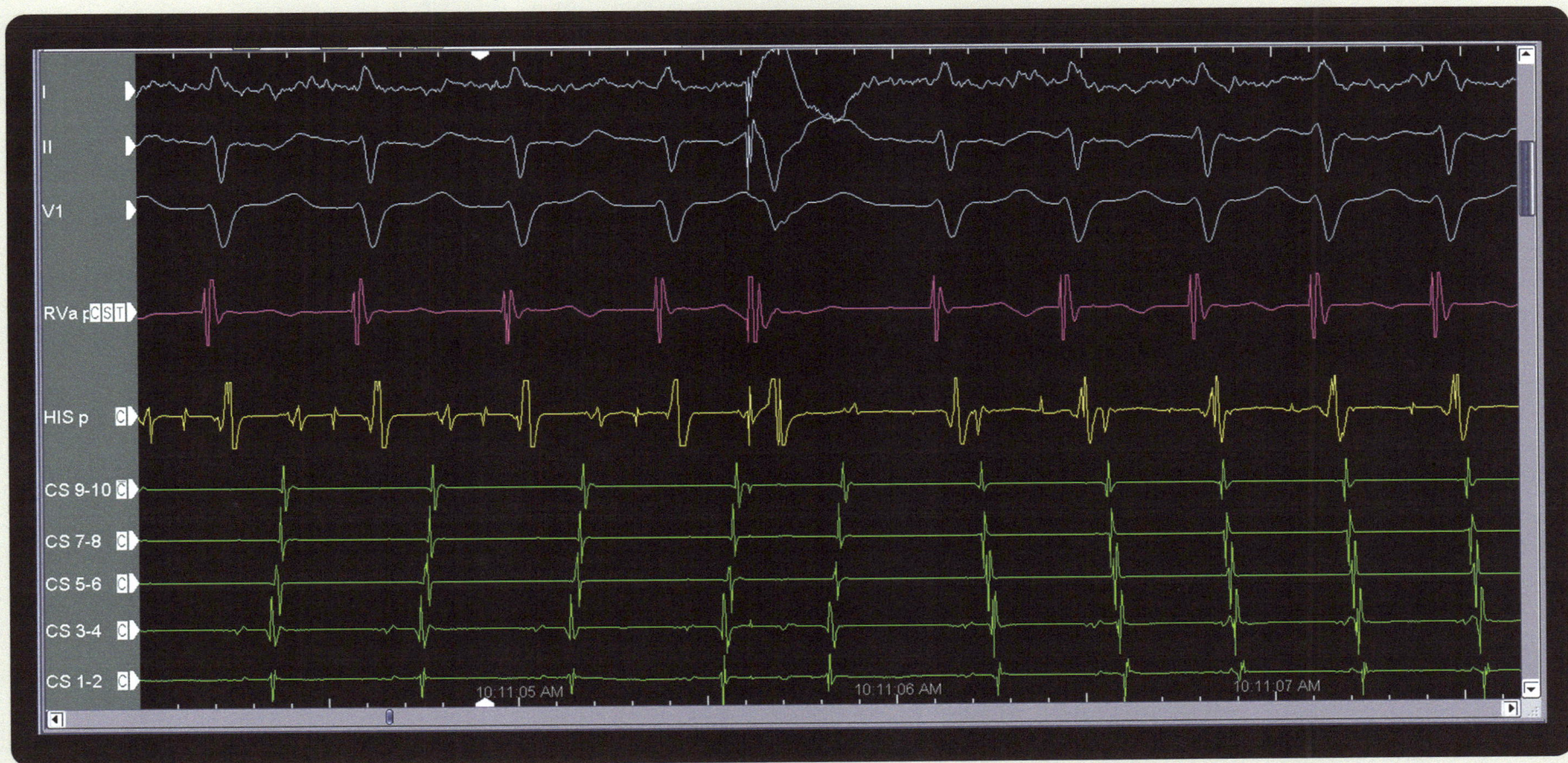

Figure 1

191

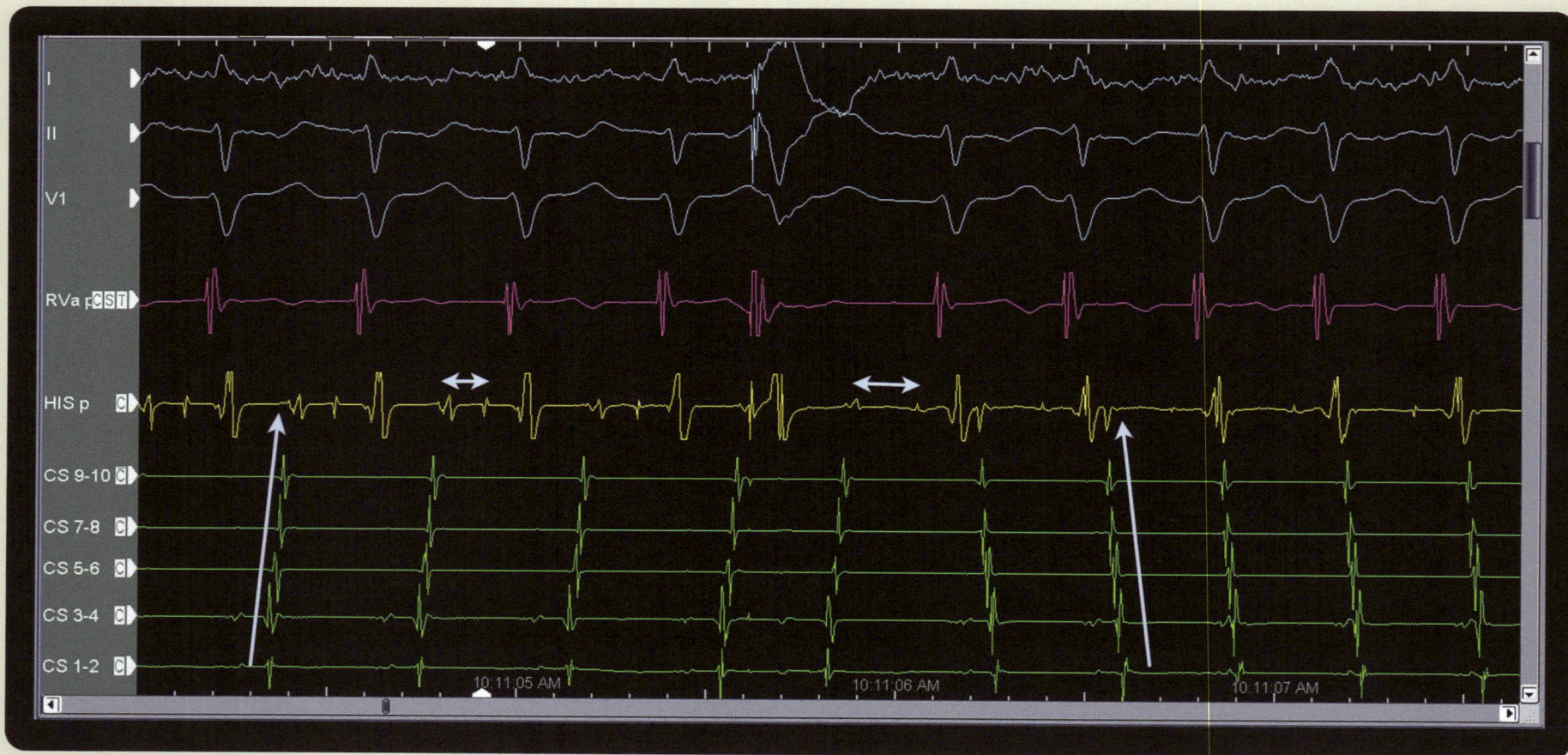

Figure 2

Discussion

At the beginning of this tracing, the tachycardia has an eccentric atrial activation pattern consistent with orthodromic AVRT using a left lateral AP. Following the stimulated PVC, the tachycardia converts to an AVNRT with a central atrial activation!

Try to explain the mechanism of how this transition occurred.

Figure 2 gives you a hint. During AVRT, the tachycardia circuit involves the antegrade fast pathway in the AV node. The induced PVC conducts retrogradely to the atrium using the AP as evidenced by the same eccentric atrial activation sequence. However, there is a sudden A-H prolongation on the subsequent beat, which suggests that the antegrade limb of the circuit has changed to a previously unrecognized slow pathway of the AV node. The most plausible explanation for the fast pathway to be refractory at this time is retrograde concealed conduction into the AV node from the induced PVC.

After antegrade conduction via the AV nodal slow pathway, there are 2 available routes of retrograde conduction back to the atria: 1) a recovered AV nodal fast pathway *or* 2) the left lateral AP.

The reentrant circuit will utilize the shortest available path being, in this situation, the retrograde fast pathway as opposed to the relatively distant left lateral AP. Hence, AVNRT is initiated!

5.3 Tachycardias 3

This is another example of unexpected findings during routine EP maneuvers! Review both tracings to discover the unusual finding!

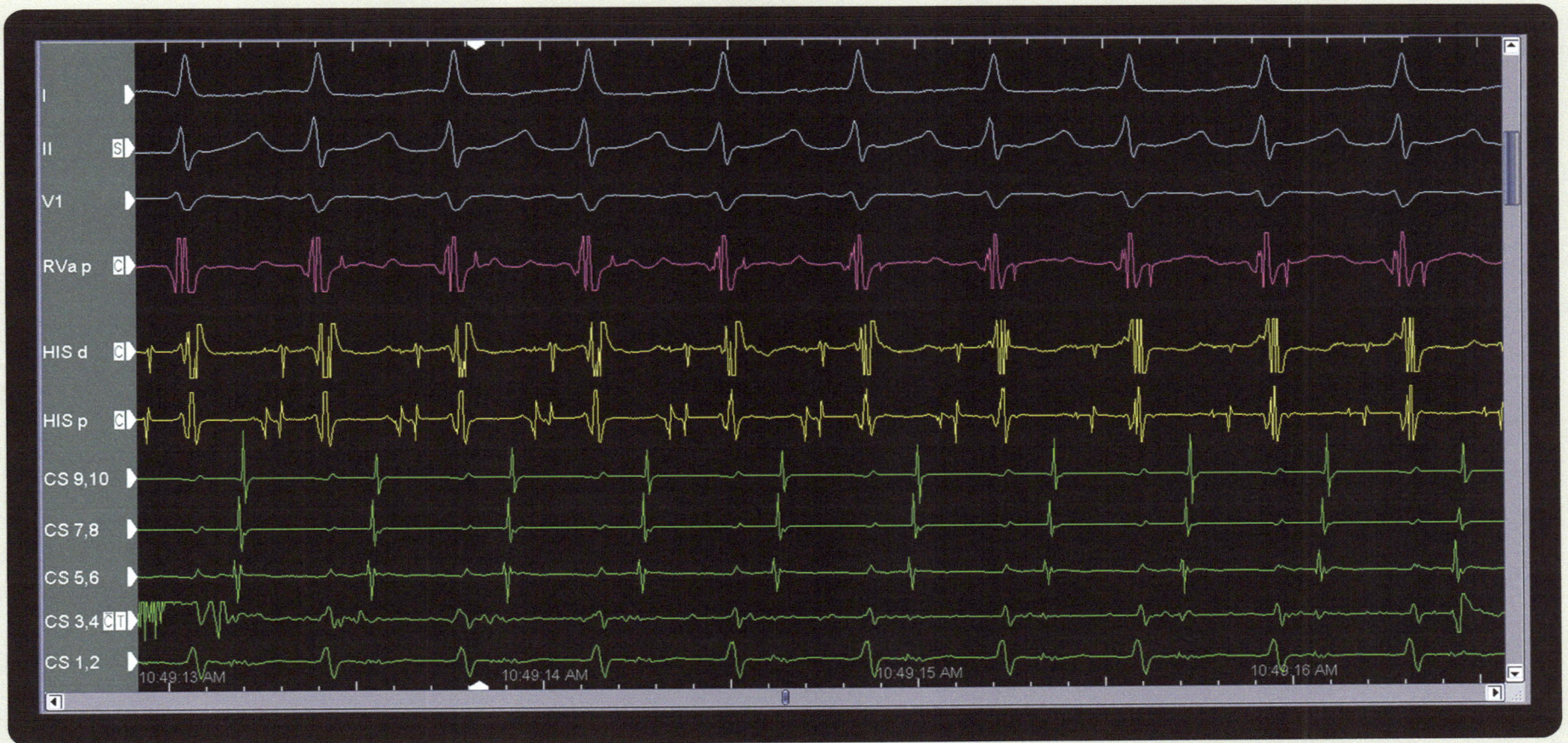

Figure 1

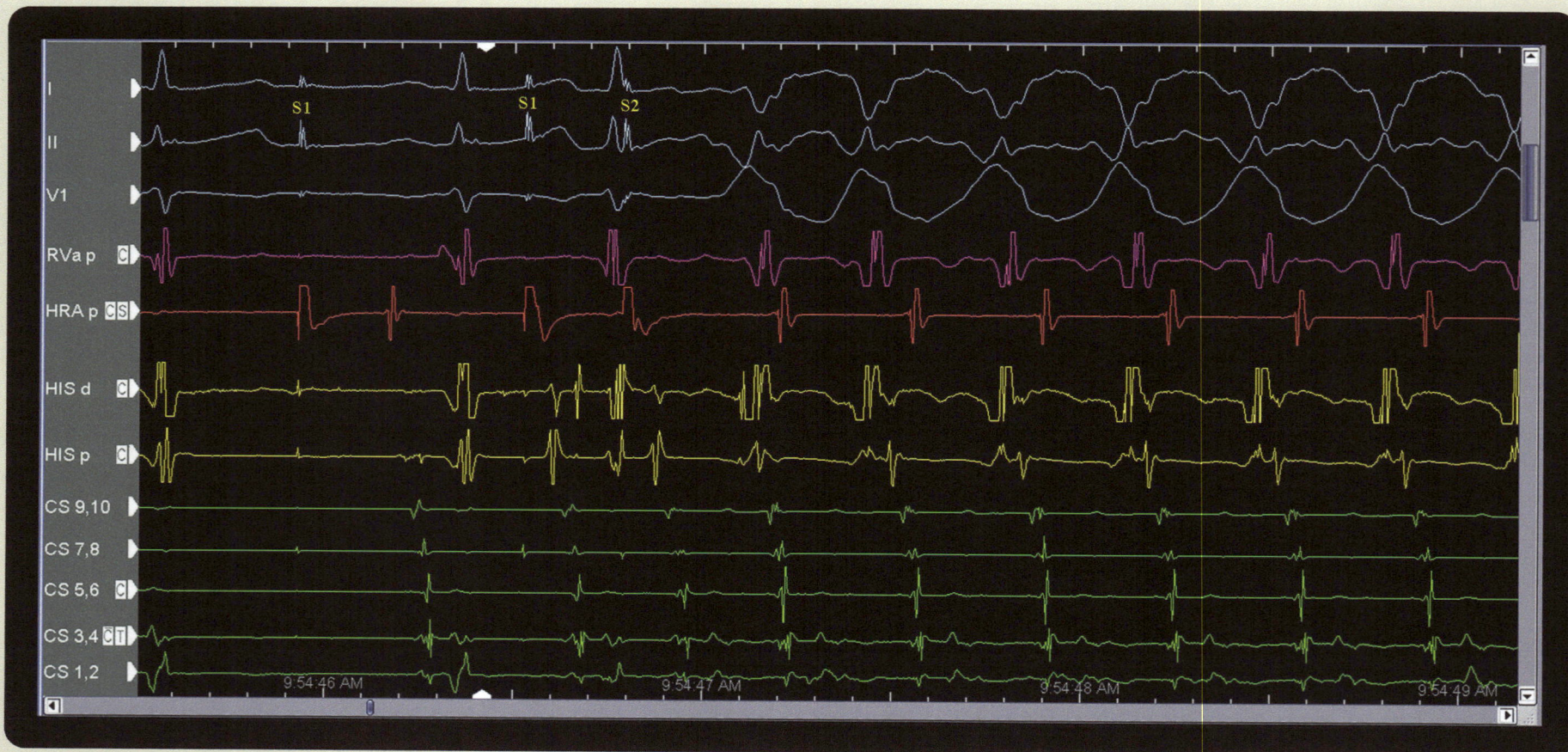

Figure 2

Discussion

Figure 1 demonstrates orthodromic AVRT. By definition, orthodromic AVRT involves antegrade conduction using the AV node and retrograde conduction using an AP, which in this tracing is a left lateral AP.

Figure 2 shows the initiation of antidromic AVRT in the same patient. By definition, antidromic AVRT involves antegrade conduction using the AP and retrograde conduction using the AV node. Try to explain the mechanism of how the tachycardia was induced in this example.

The first S1 seen in this tracing fails to capture the atrium and is followed by a native sinus beat, which conducts to the ventricle using the AV node (as evidenced by the narrow QRS on the surface leads). The second displayed S1 captures the atrium and subsequently conducts antegradely to the ventricles using the AV node. The S2 captures the atrium and also conducts to the ventricle, but the QRS morphology has changed. This is due to conduction block in the AV node and conduction to the ventricles exclusively via a left lateral AP.

We know it is a left lateral AP since the surface ECG demonstrates marked right axis deviation and a RBBB-like morphology. As well, the earliest "V" electrogram can be seen in CS 1-2. The tachycardia circuit continues from the ventricle to the atrium retrogradely via the AV node (notice that the earliest atrial activation is in the His channel).

The induction of both antidromic and orthodromic AVRT in the same patient is quite unusual!

5.4 VT

This tracing demonstrates ventricular tachycardia (VT) initiation using ventricular extra-stimulus pacing. This is an example of a very specific type of VT. What is it?

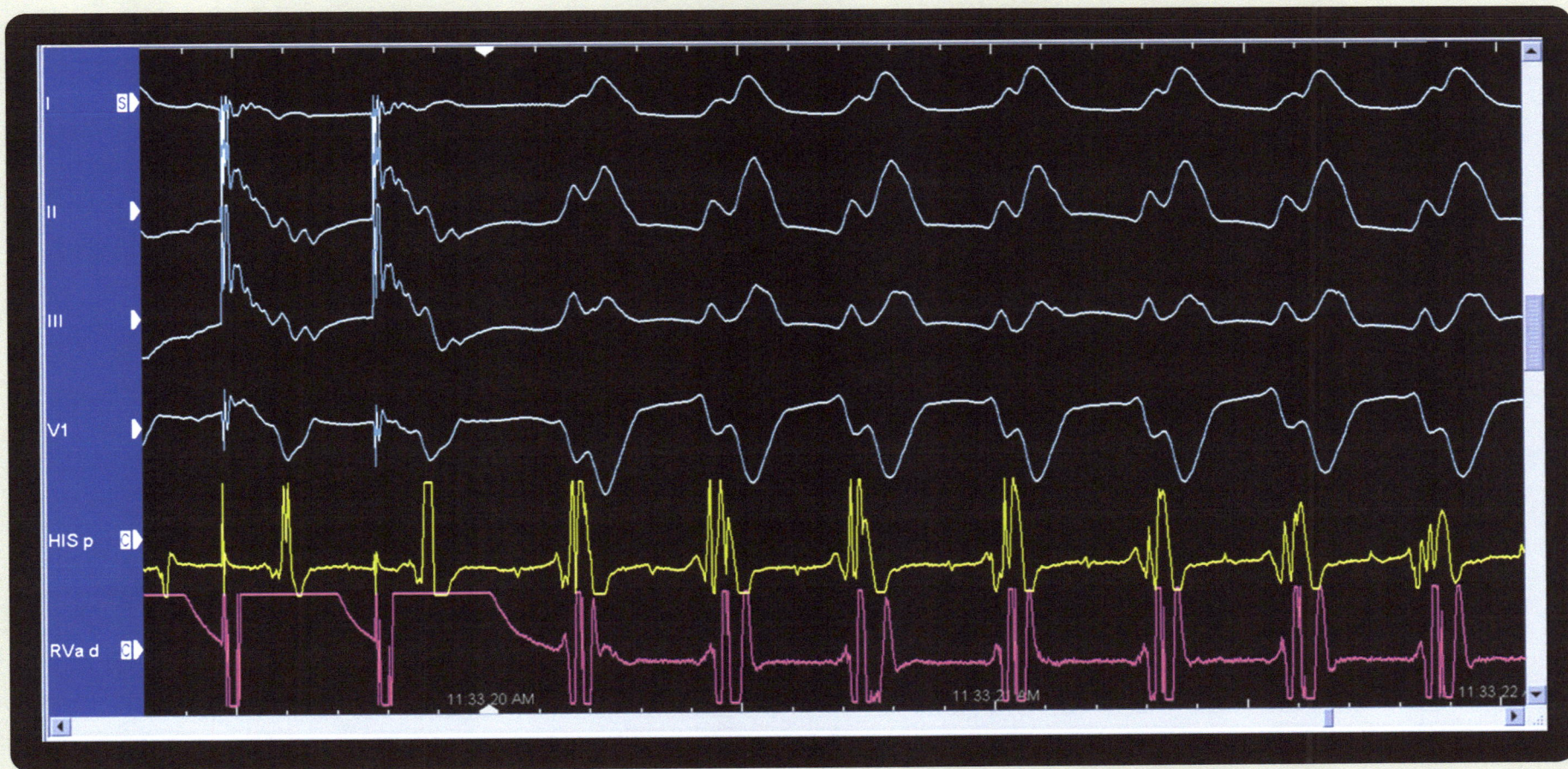

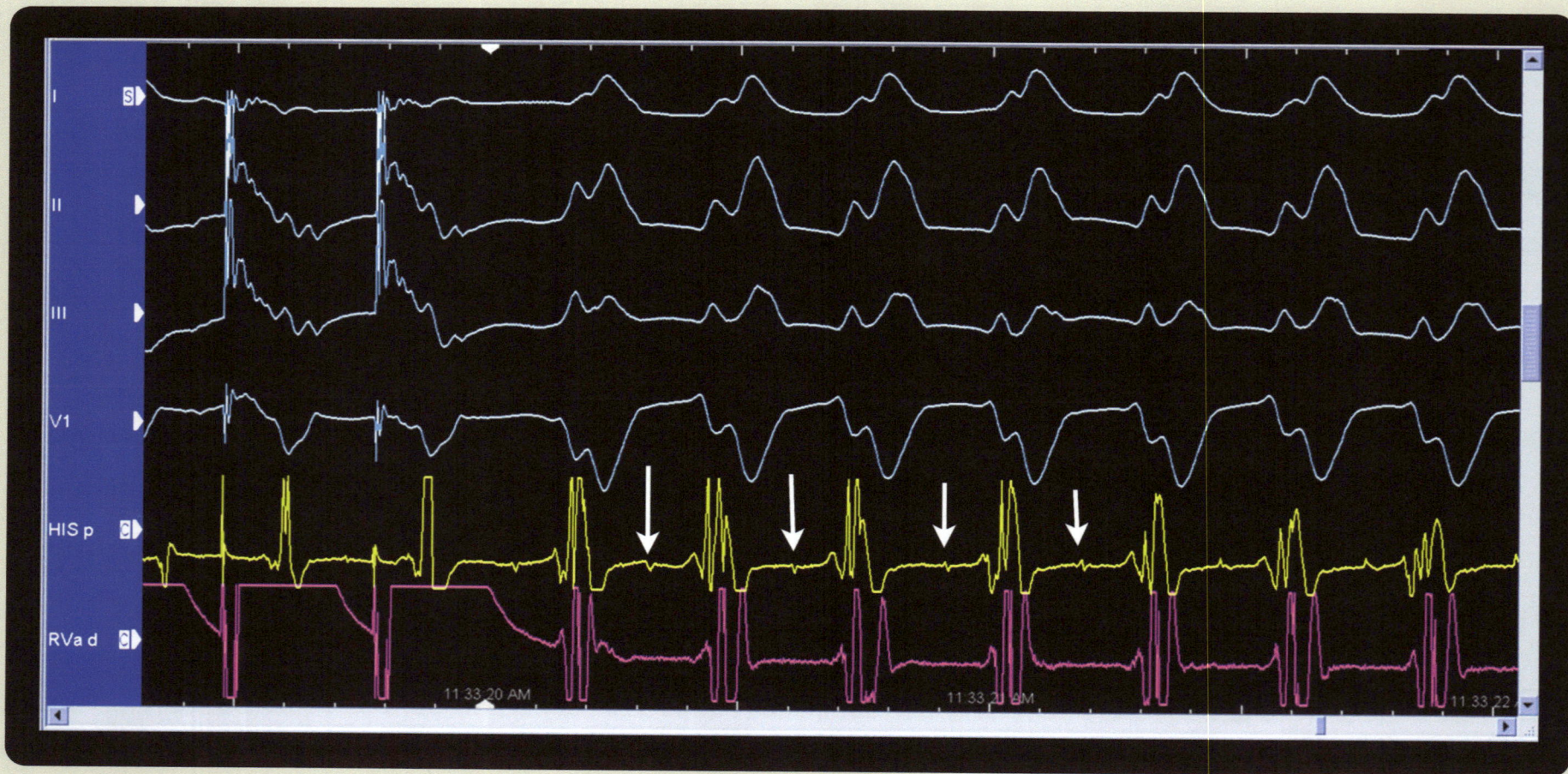

Discussion

There is a differential diagnosis for a wide-complex tachycardia induced with ventricular extra-stimulus pacing including: 1) VT; 2) SVT with aberrancy *or*; 3) a tachycardia using an AP (preexcited tachycardia).

While the exact diagnosis cannot be confirmed on this tracing alone, the diagnosis was VT and, more specifically, bundle branch reentry VT. Notice that there is a His catheter in place and that the His electrograms (*white arrows*) have a 1:1 relationship with the ventricular electrograms.

The typical features of bundle branch reentry VT include: 1) a typical LBBB morphology on the surface ECG; 2) a His EGM in a 1:1 relationship with the ventricle and; 3) a relatively long H-V interval.

The mechanism of bundle branch reentry VT is a macro-reentrant circuit using the His-Purkinje system.

Most commonly, the circuit involves antegrade conduction down the right bundle (therefore a LBBB morphology), transseptal myocardial conduction, followed by retrograde conduction using the left bundle branch. The upper common pathway of this circuit is the His bundle. Therefore, a retrograde His, in a 1:1 relationship with the VT, is the critical diagnostic finding confirming bundle branch reentry.

5.5 Substrate for?

This is a good example of *not* making assumptions about a patient's tachycardia mechanism!

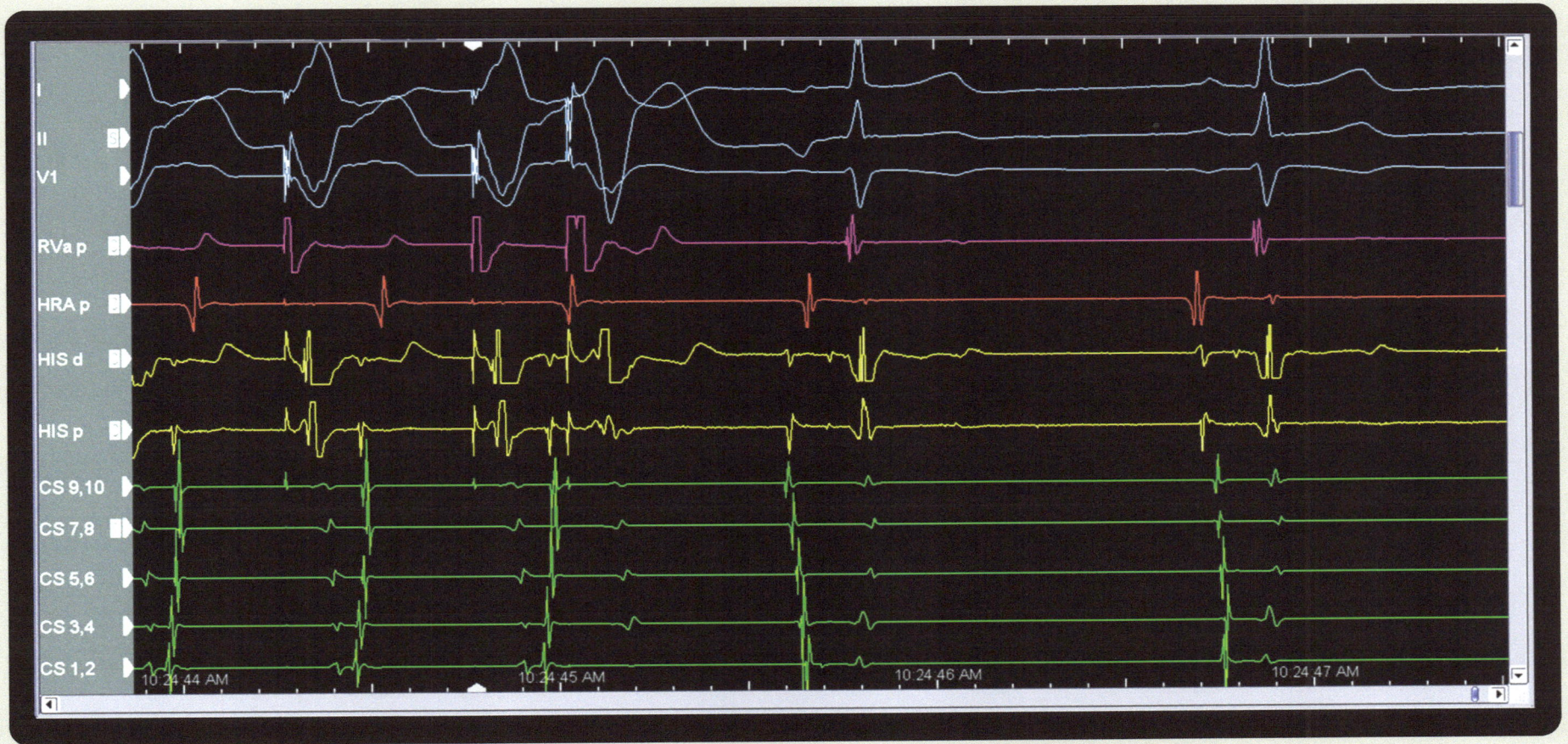

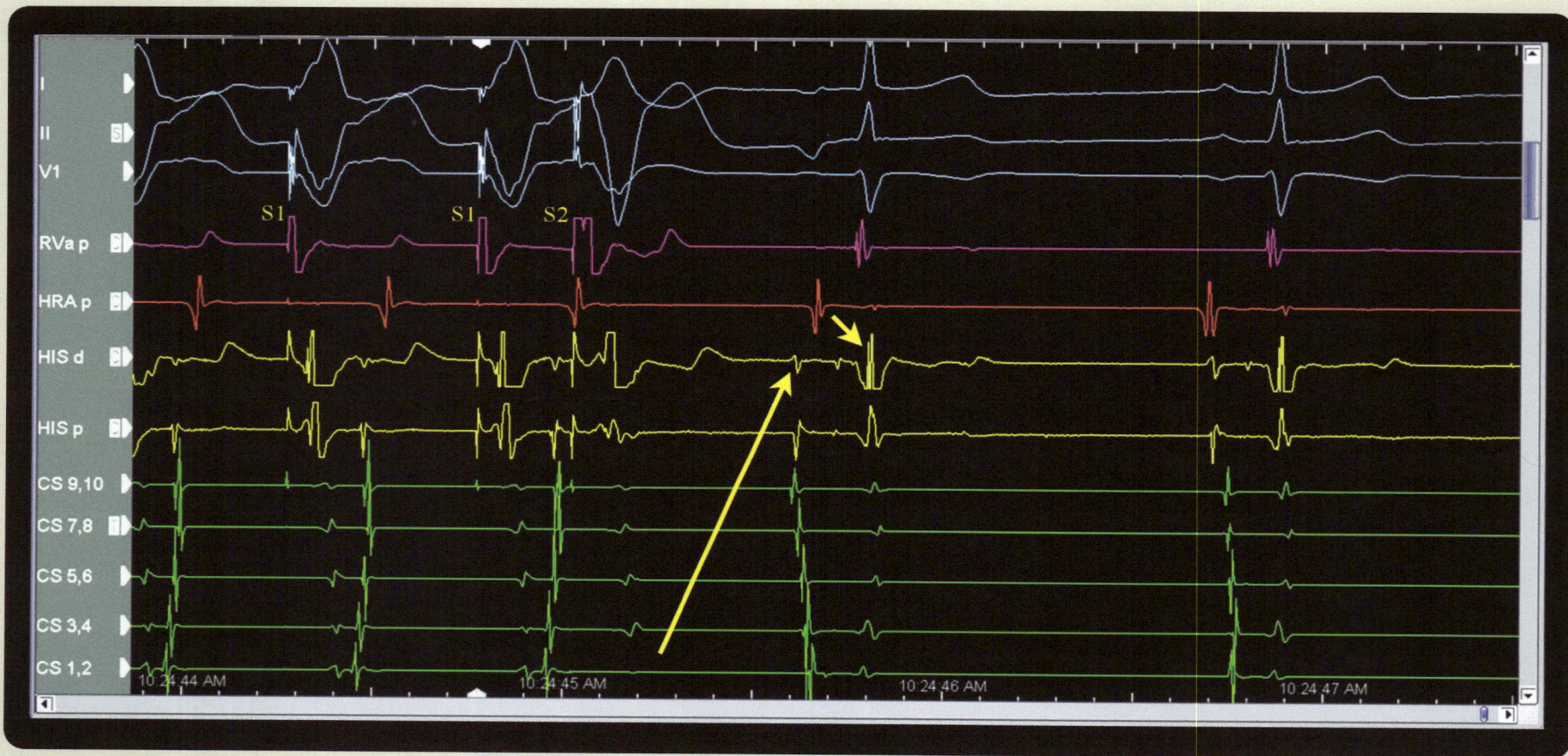

Discussion

This tracing demonstrates ventricular extra-stimulus pacing. Notice that the 2 S1's conduct to the atria with an eccentric atrial activation pattern using a left-lateral AP. One would correctly assume that this patient has the substrate for orthodromic AVRT.

However, notice the response to the ventricular S2. It *appears* to block and is followed by an atrial depolarization with a His "A" first.

The differential diagnosis for this observation include:

1. The S2 truly fails to conduct retrogradely to the atrium (thus we have reached the ERP of the AP and the AV node) and is followed by a PAC, which conducts to the ventricle.

2. The S2 conducts to the atrium via a retrograde slow AV nodal pathway (*long yellow arrow*). This is followed by conduction antegradely via a fast pathway to the ventricle (*short yellow arrow*).

The S1S2 maneuver was repeated several times with a reproducible result. This speaks well for scenario #2 rather than scenario #1.

The presence of an atypical AV nodal echo beat (retrograde conduction via the slow pathway and antegrade conduction via the fast pathway) is an indication that the patient also has the substrate for atypical AVNRT. In fact, this patient demonstrated both AVRT and atypical AVNRT!

5.6 12-lead Diagnosis

This is another example of why being alert and focused during catheter insertion can be instructive! Ectopy which is often induced during catheter manipulation may reveal the diagnosis before the formal EP study even begins!

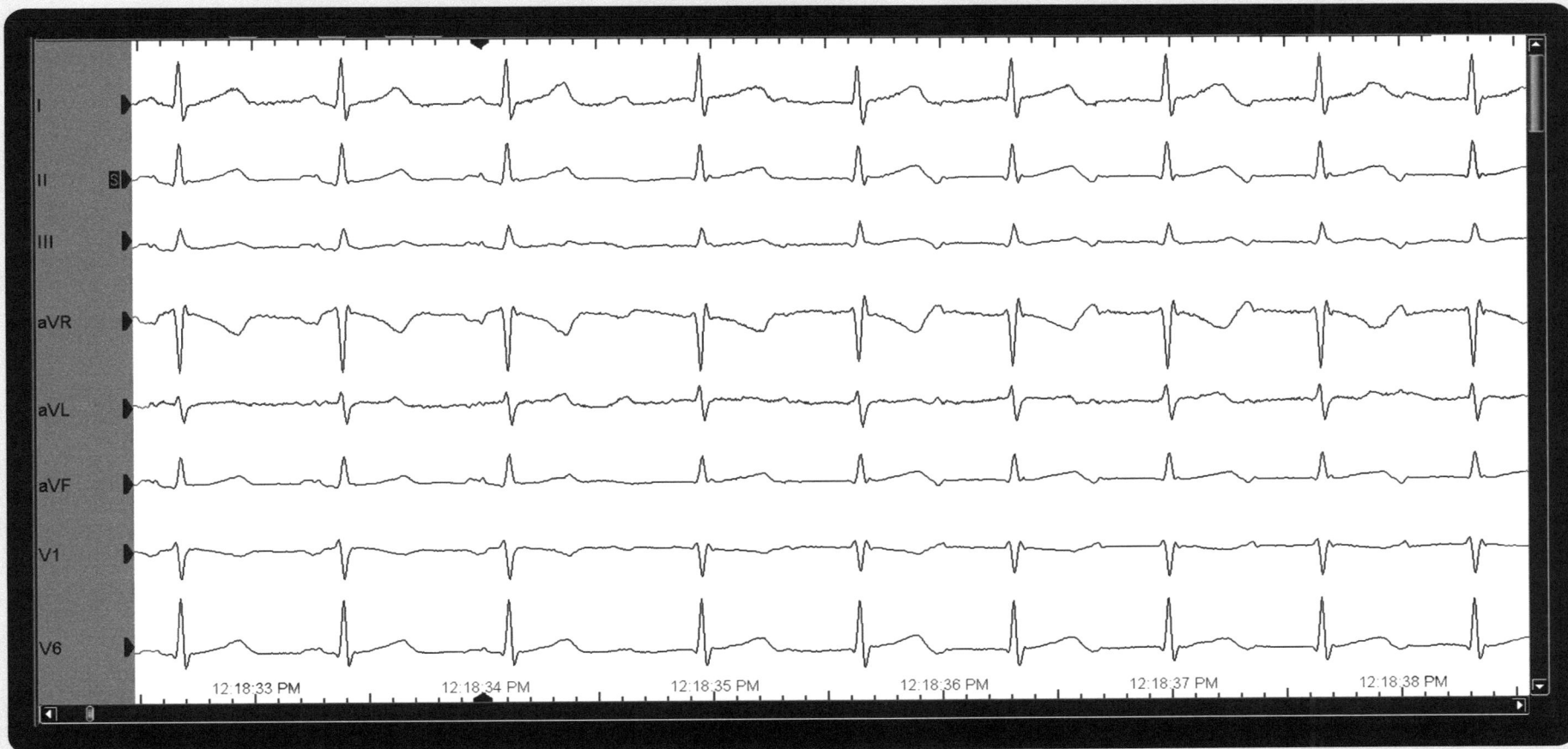

Figure 1

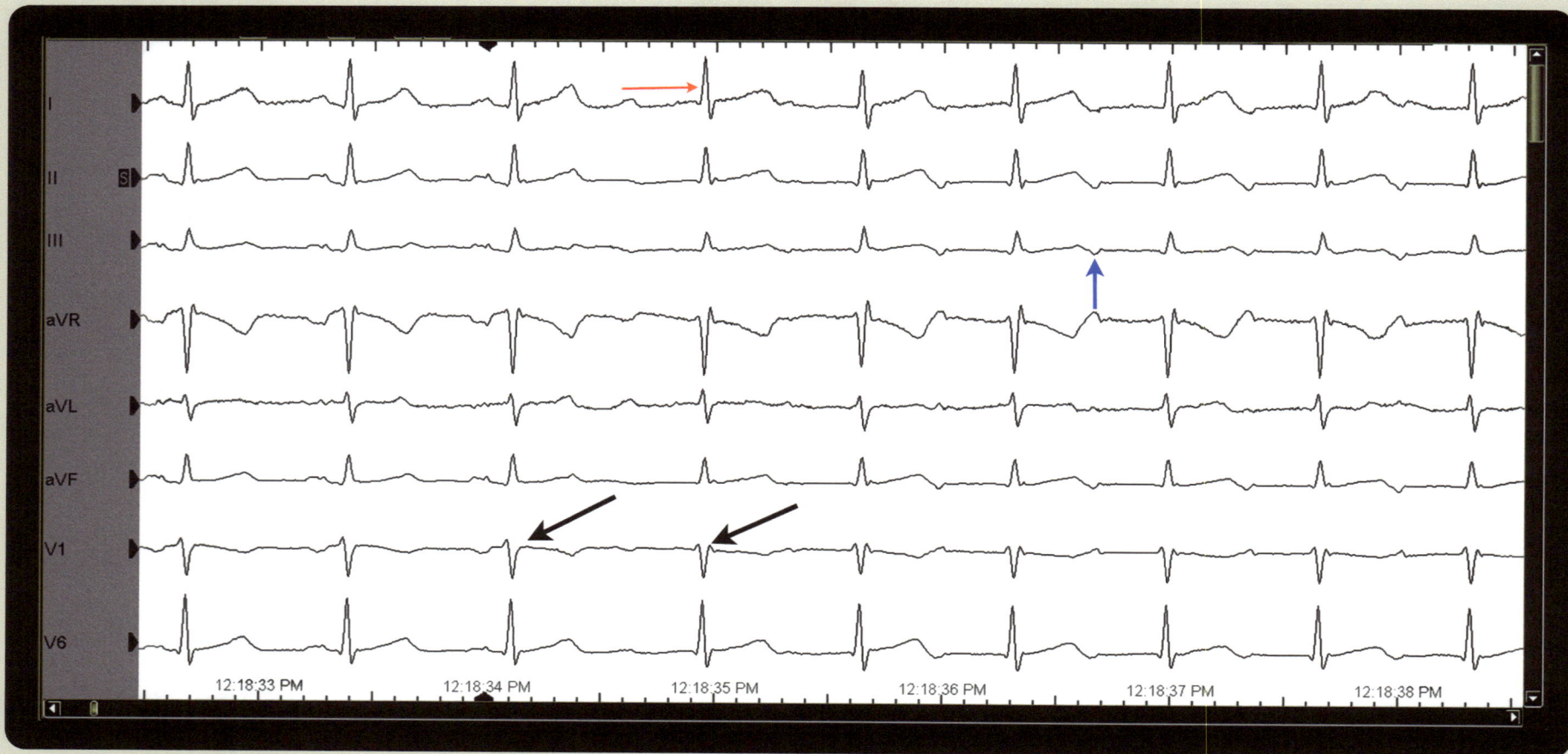

Figure 2

Discussion

Figure 1 shows sinus rhythm and then the onset of a narrow-complex tachycardia with the usual differential diagnosis.

Figure 2 gives you the hints to the diagnosis. The *red arrow* shows an atrial ectopic beat that causes sudden PR prolongation, likely indicating antegrade conduction via an AV nodal slow pathway. The *blue arrow* points to atrial activity during the tachycardia. The first *black arrow* draws your attention to the morphology of V1 during sinus rhythm and the second *black arrow* to the subtle change in morphology during tachycardia. This is suspicious for a second P wave.

continues

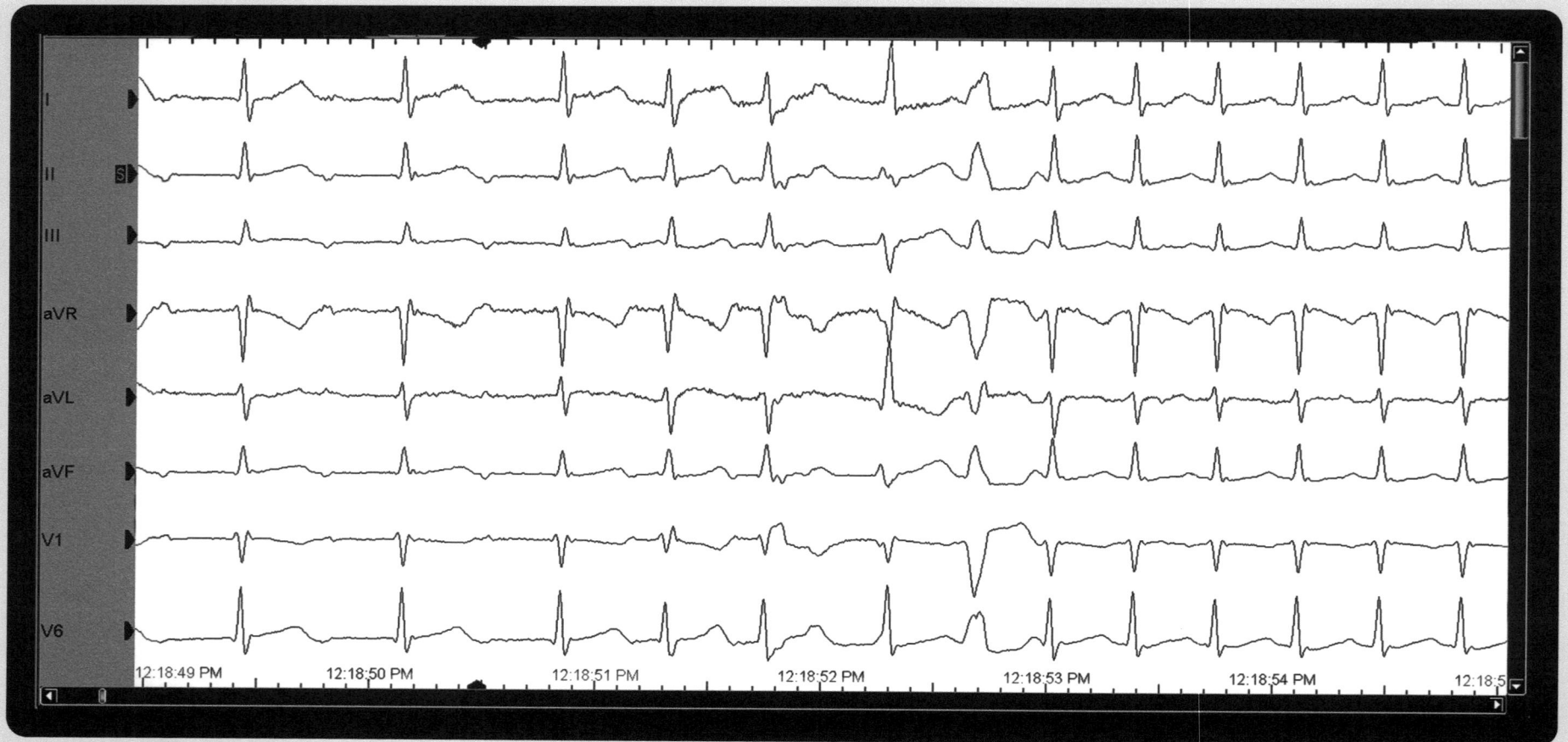

Figure 3

Figure 3 shows more ectopy during the ongoing tachycardia and a transition to a faster tachycardia on the right half of the tracing.

continues

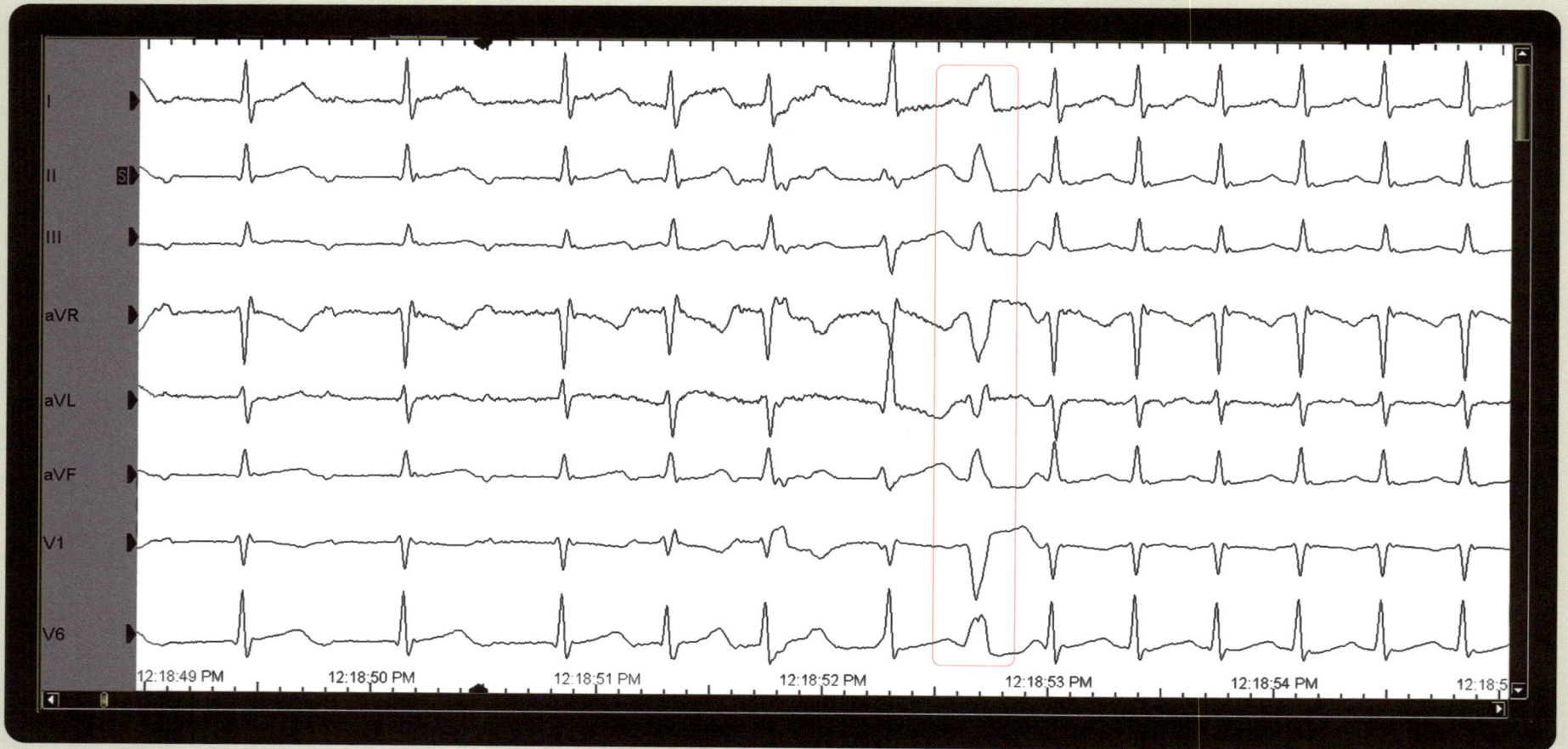

Figure 4

Figure 4 gives you yet another hint. The circled PVC precedes the increase in the tachycardia rate. Notice that the P waves in the slower tachycardia are at the midpoint between the QRSs. Also notice that the faster tachycardia following the PVC is exactly half the cycle length (twice the rate) of the tachycardia prior to the PVC.

This is an example of AVNRT with 2:1 conduction to the ventricle. The subsequent PVC primed or "reset" the conduction system by influencing the refractory period of the His, thus allowing subsequent 1:1 conduction.

5.7 Break

This tracing demonstrates another example of concealed conduction. Explain the termination of the tachycardia.

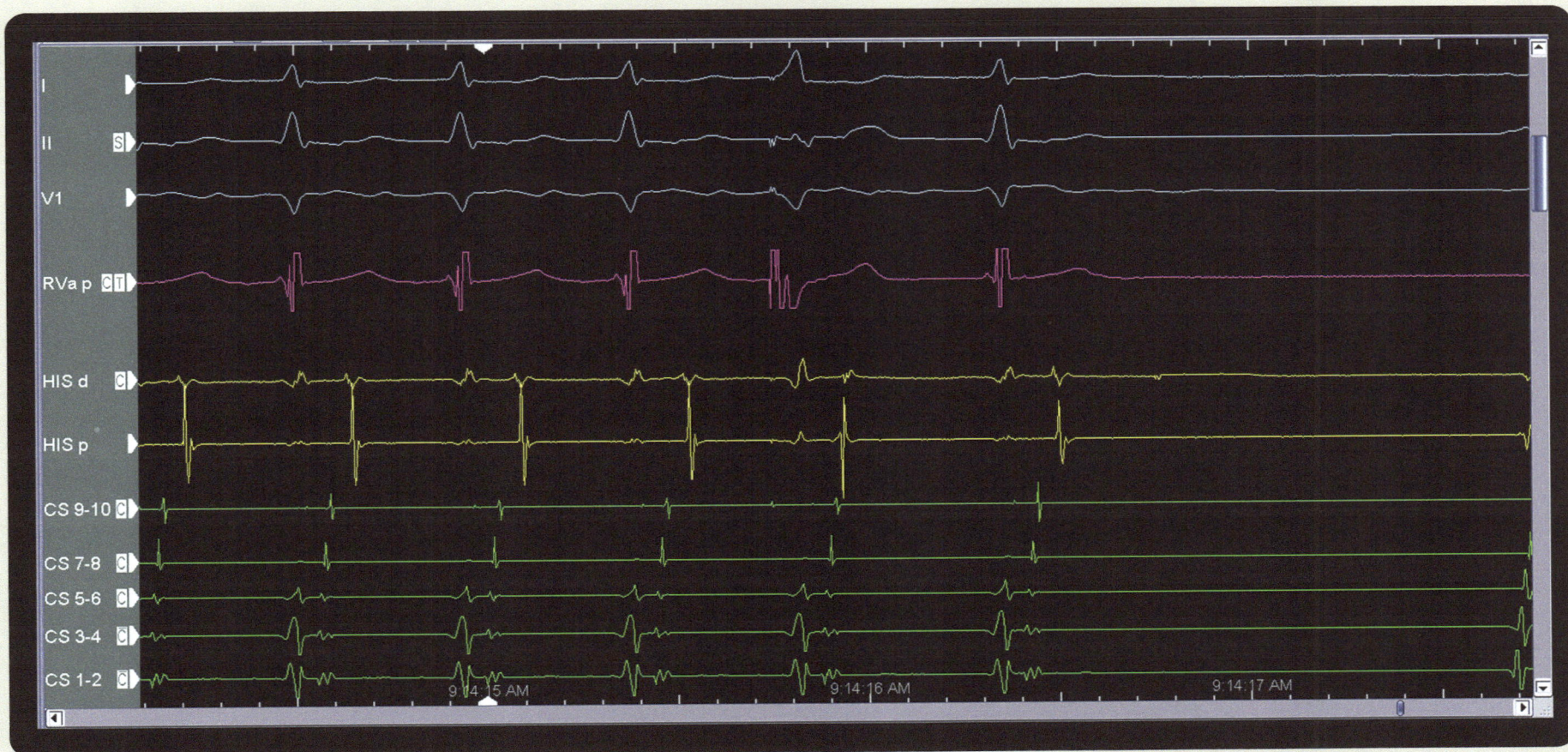

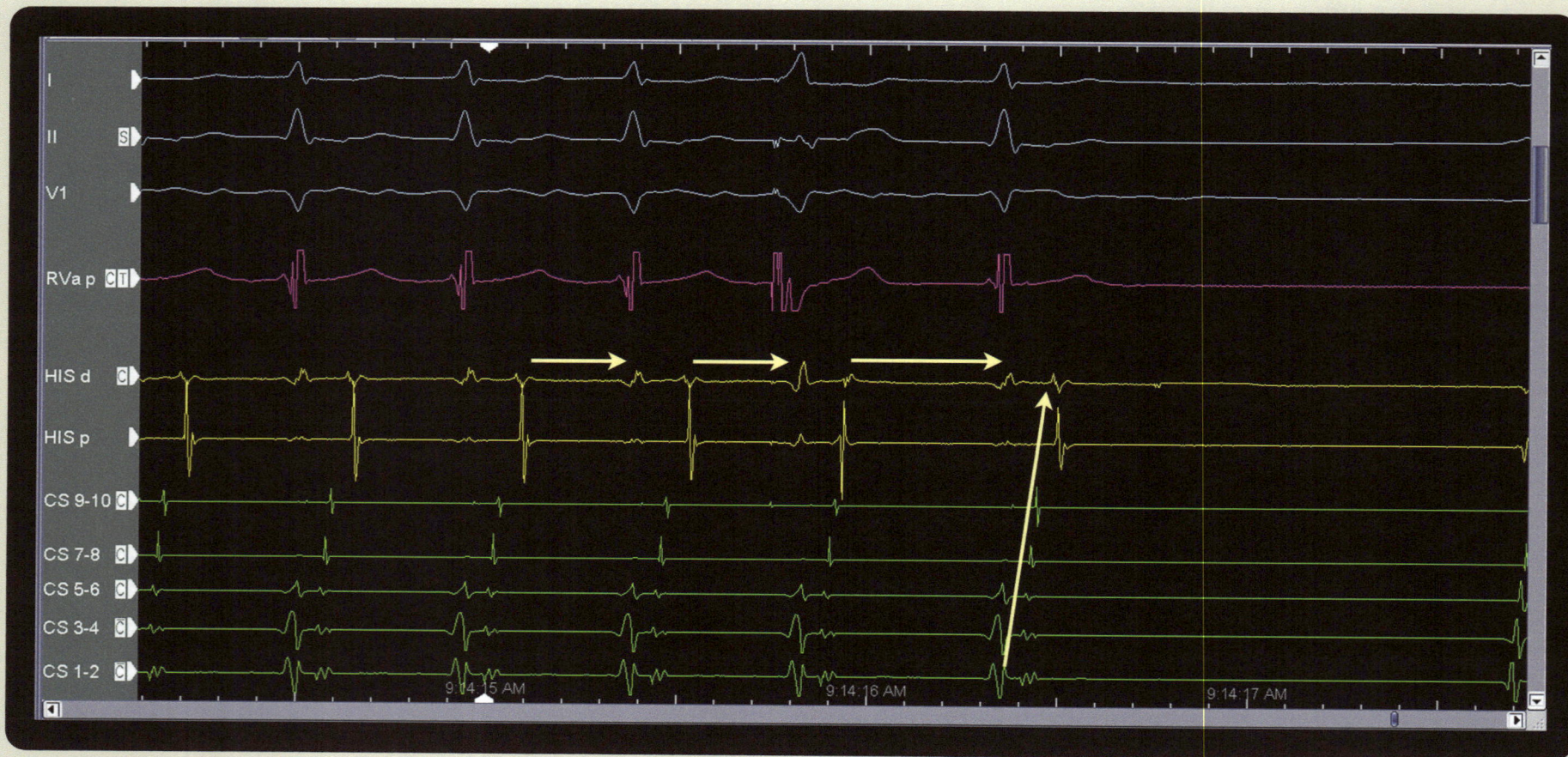

Discussion

The tracing begins with a 1:1 narrow-complex tachycardia with an eccentric atrial activation pattern (earliest atrial activity in CS 3-4) that has a differential diagnosis of either AVRT using a left lateral AP *or* a left atrial tachycardia.

A PVC is introduced during the tachycardia at a time when the His is refractory. In a normal conduction system, a His-refractory PVC cannot conduct retrogradely to the atrium since the AV node is refractory. However, when an AP exists, an alternate pathway to the atrium is available. In this example, the His-refractory PVC is delivered, which shortens the subsequent atrial cycle length. By doing so, the existence of a left lateral AP is proven making AVRT the most likely diagnosis.

Even more interesting is attempting to explain how the PVC led to the termination of the AVRT. Note that the A-V interval prolongs (*yellow arrows*) on the beat following the PVC. This suggests that the AV nodal fast pathway's refractory period was reached and antegrade conduction continued using a previously undiagnosed AV nodal slow pathway.

The circuit continued with retrograde conduction via the AP but the tachycardia terminated.

To maintain the tachycardia, the circuit previously required antegrade conduction via the AV nodal fast pathway. The fast pathway should have had sufficient time to recover since the retrograde portion of the circuit is a rather distant left lateral AP, but in fact failed to conduct and the AVRT terminated. Therefore, the fast pathway must still be refractory due to retrograde concealed conduction into it after antegrade slow pathway conduction.

Notes

Cardiac Electrophysiology 2

CHAPTER 6 CASE STUDIES

6.1 Case 1

Here is a series of tracings presented in the order
that they were performed. Read through each tracing and
establish the diagnosis.

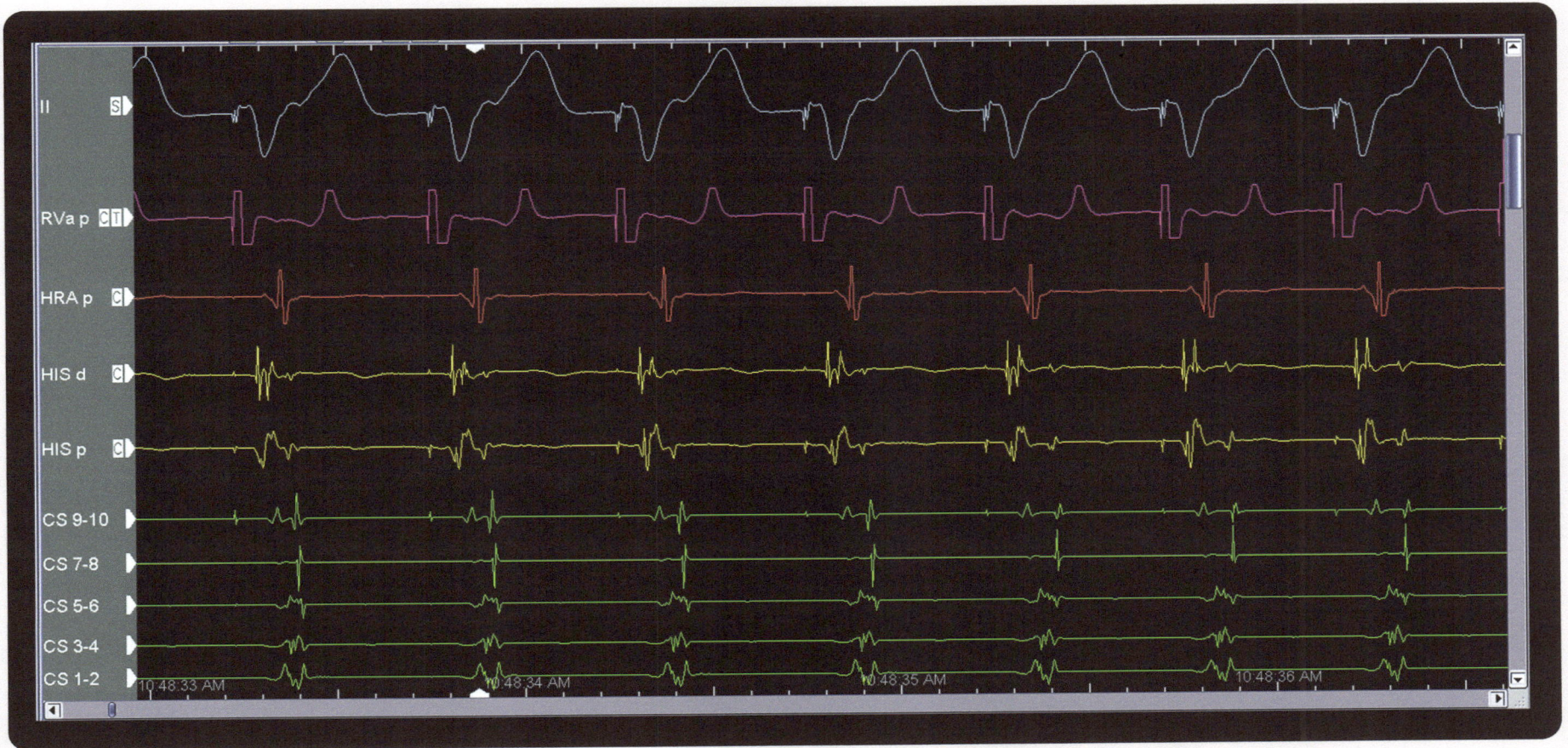

Figure 1

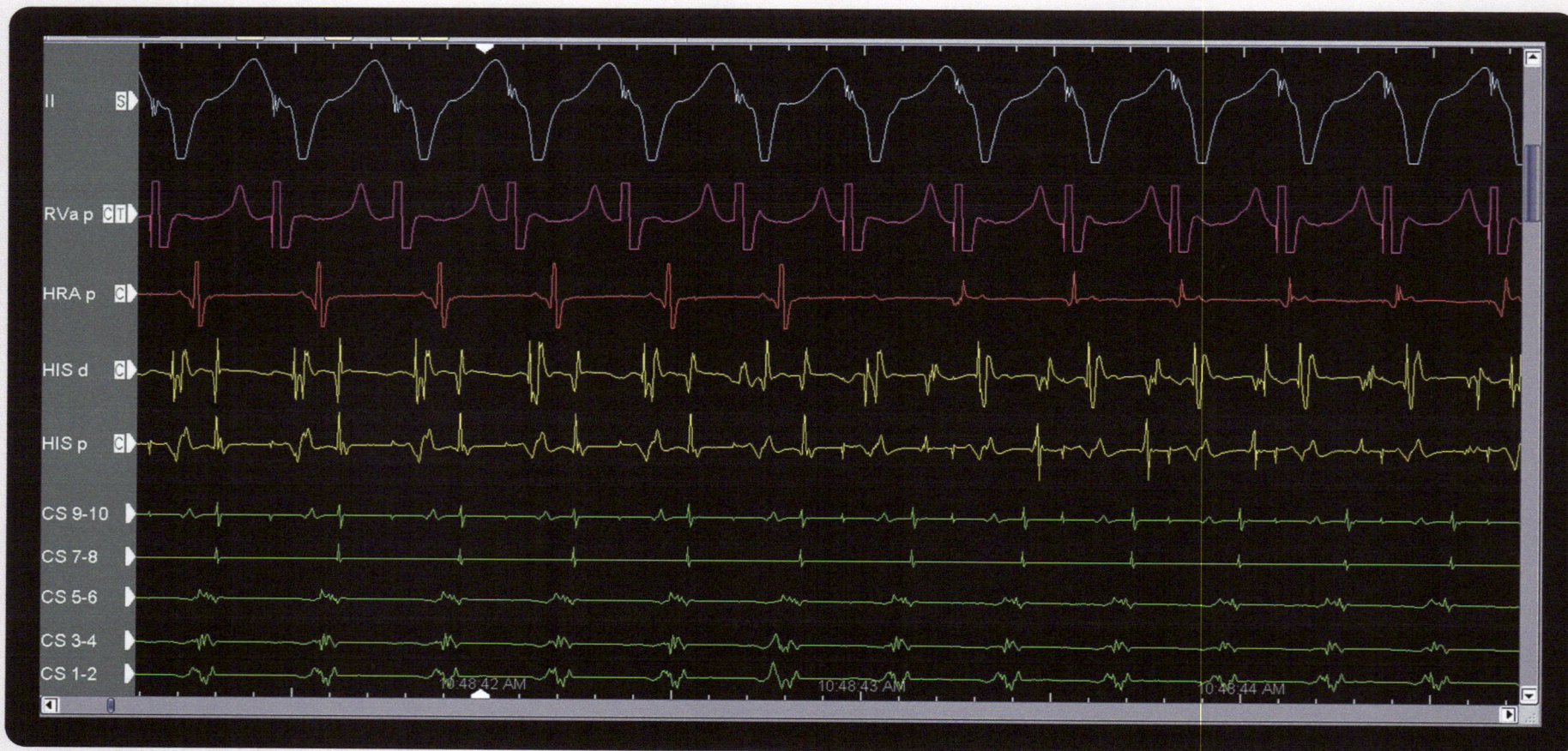

Figure 2

Discussion

Figure 1: During incremental "V" pacing, the His "A" delays, leaving CS 3-4 as the earliest "A." The timing of the HRA "A" should also be noted.

Figure 2: Nearing the completion of incremental "V" pacing, note the sudden change in position of the HRA "A" electrogram. This implies block in a right-sided AP.

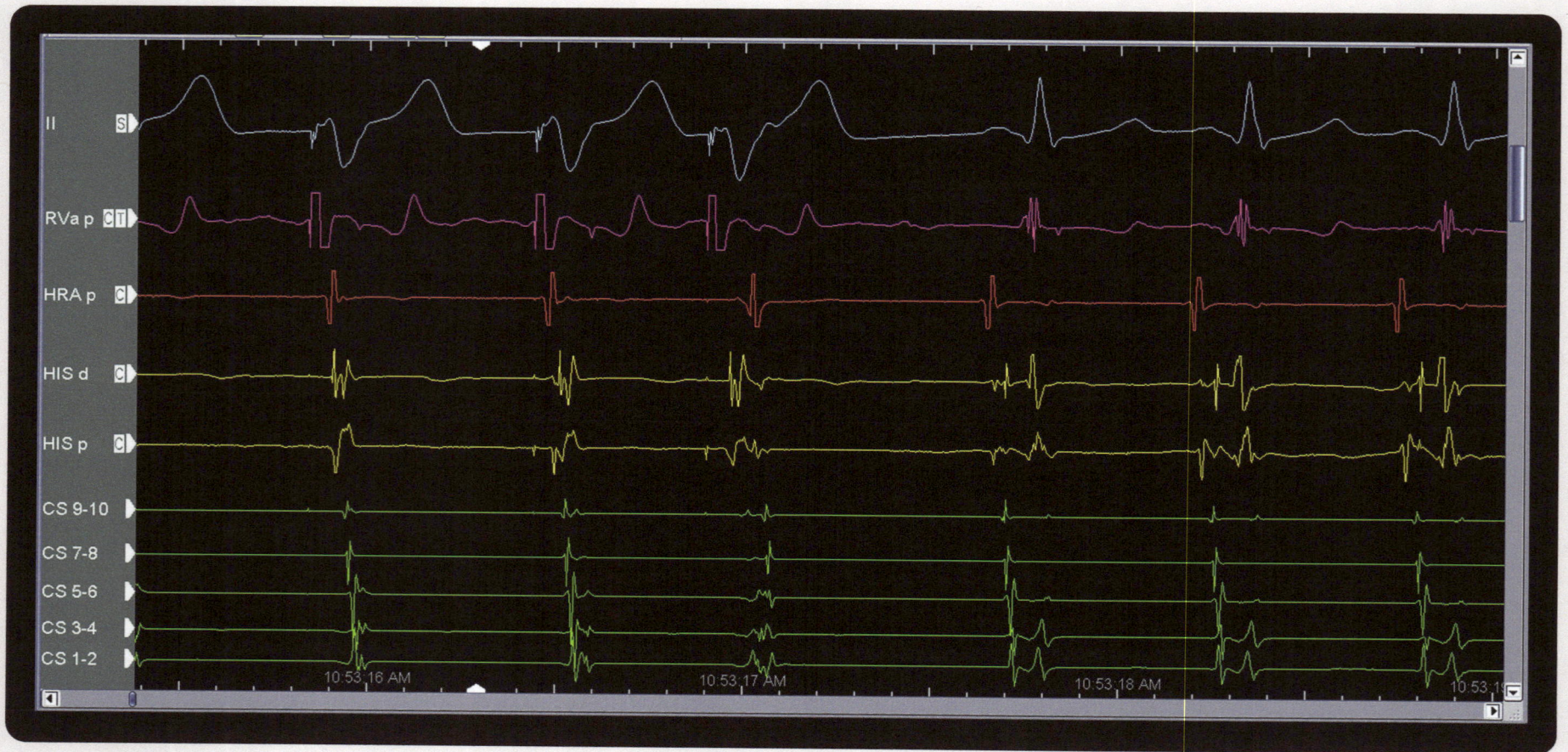

Figure 3

Figure 3: This is the beginning of "V" extras. Note the atrial activation sequence.

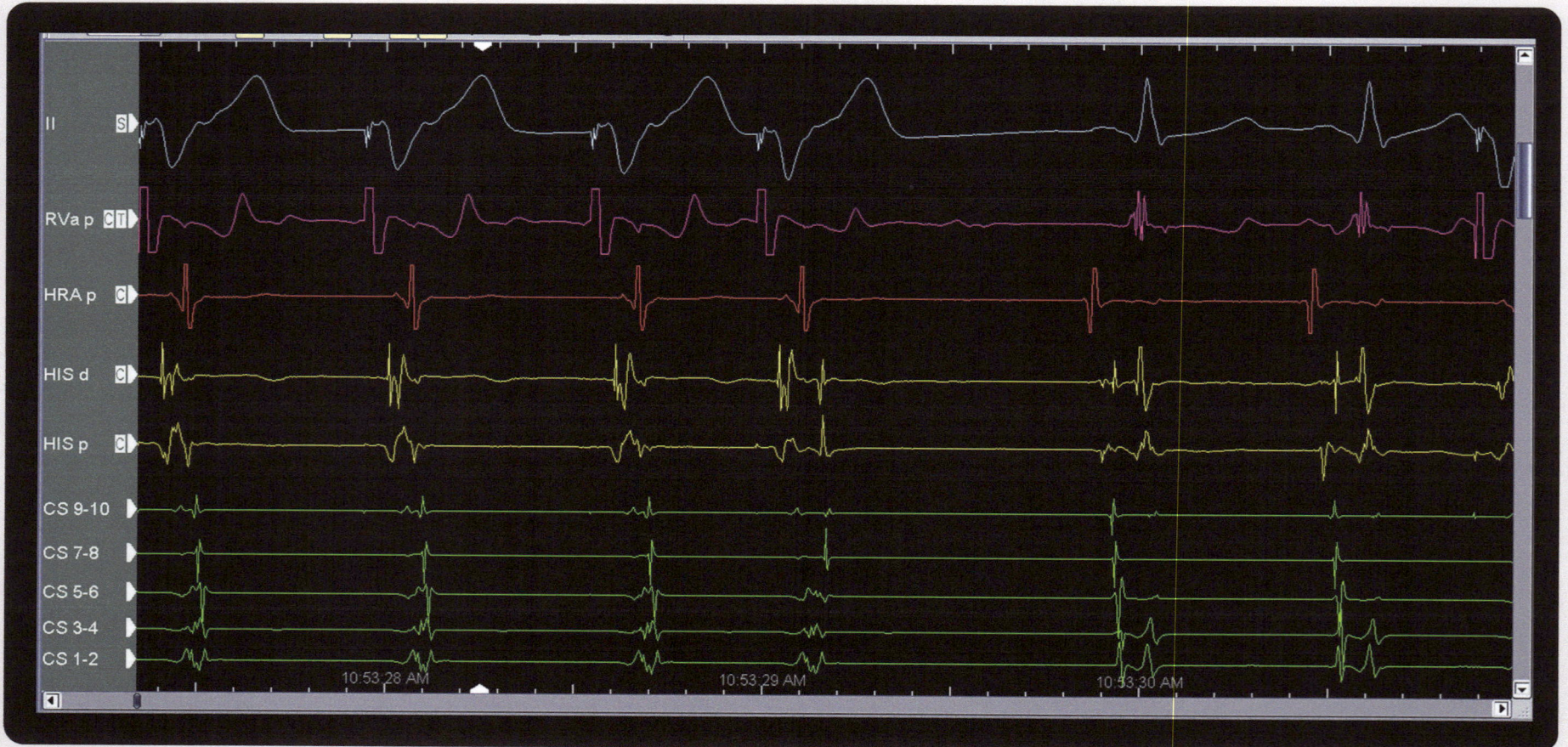

Figure 4

226

Figure 4: The His "A" has delayed leaving CS 3-4 and HRA early.
This implies AV nodal decrement.

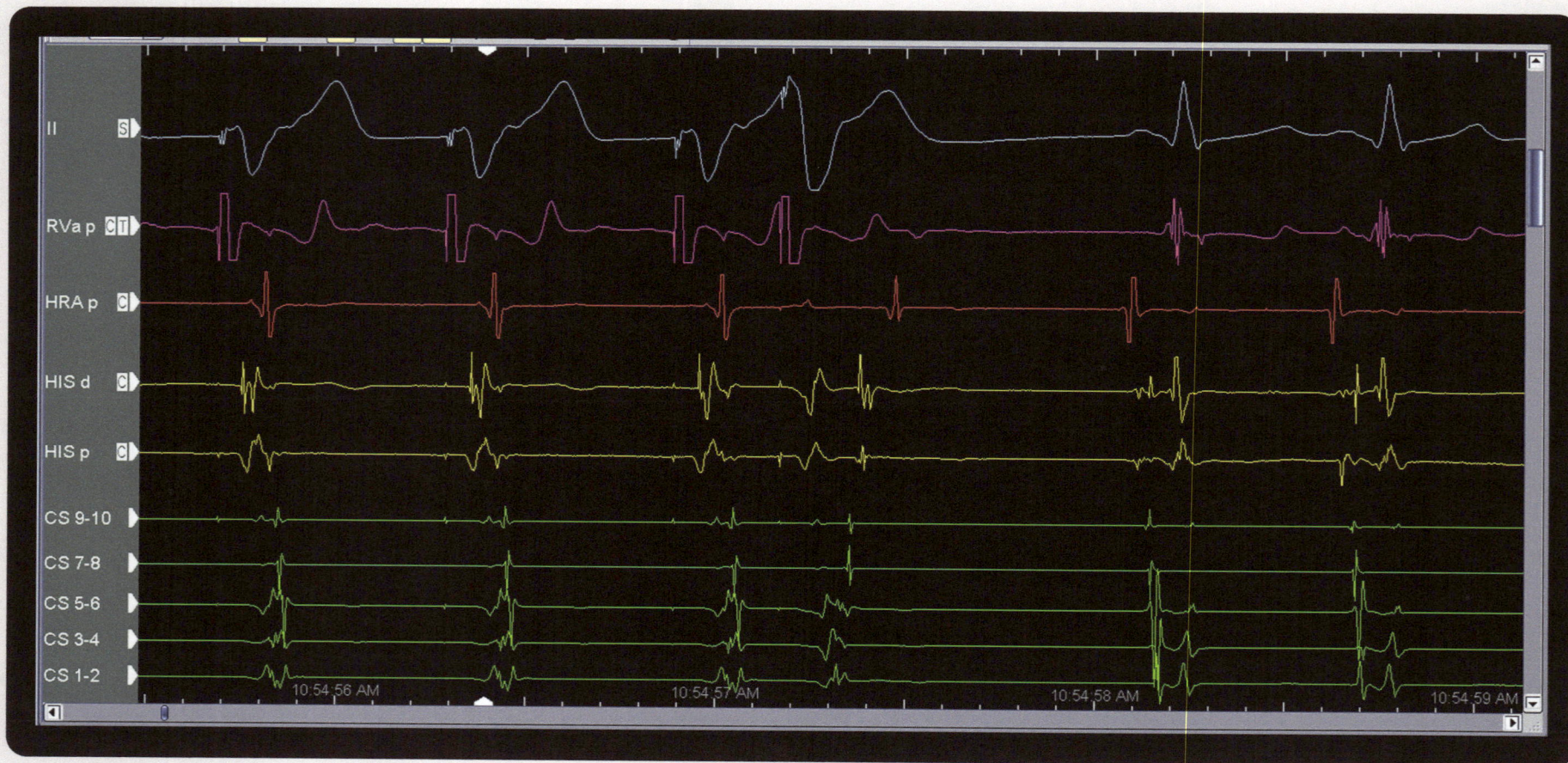

Figure 5

Figure 5: Describe what you see compared to the previous tracing. Both the His "A" and the HRA "A" have jumped late. This implies block in the right-sided pathway and the AV node.

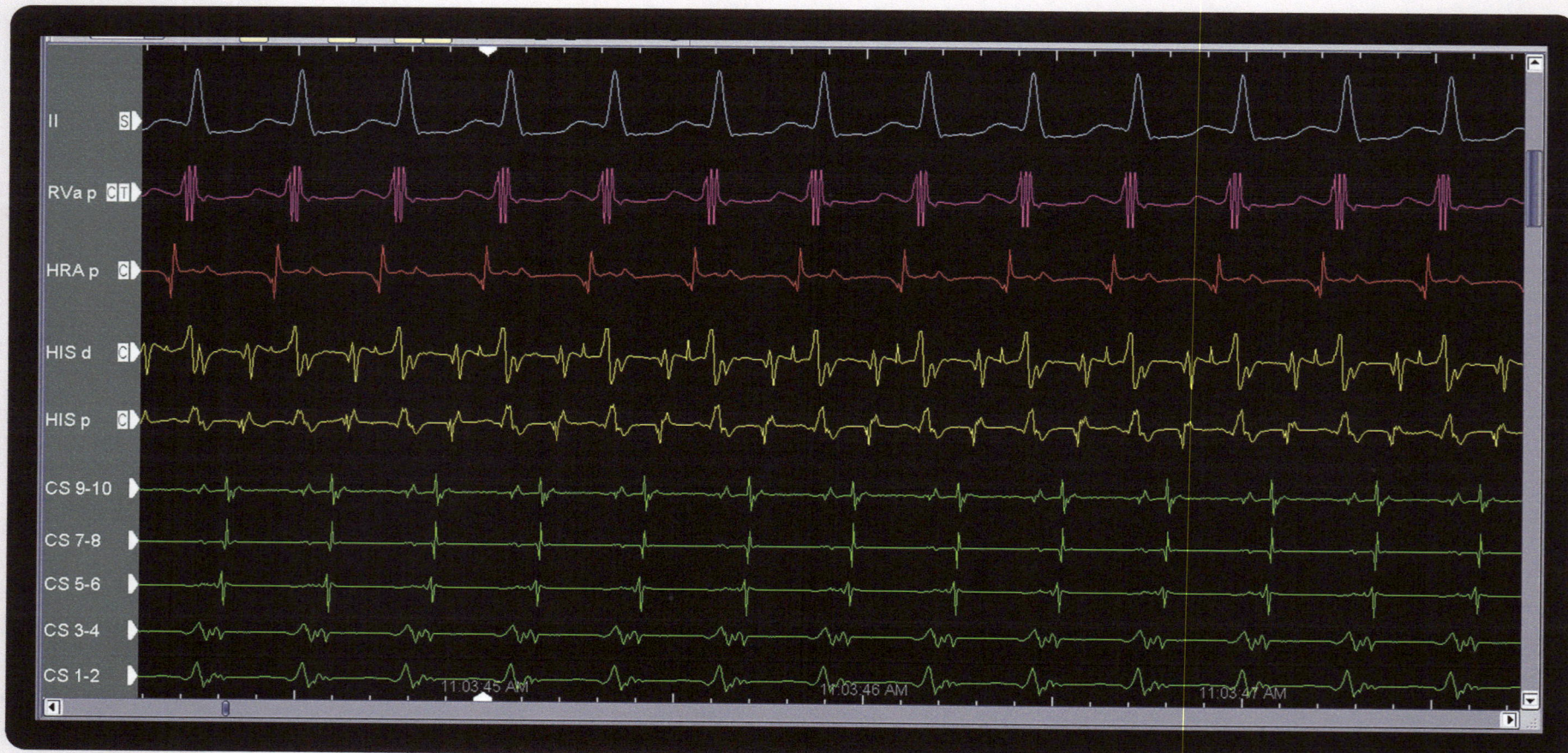

Figure 6

Figure 6: This is tachycardia 1. It is AVRT utilizing a left-lateral AP.

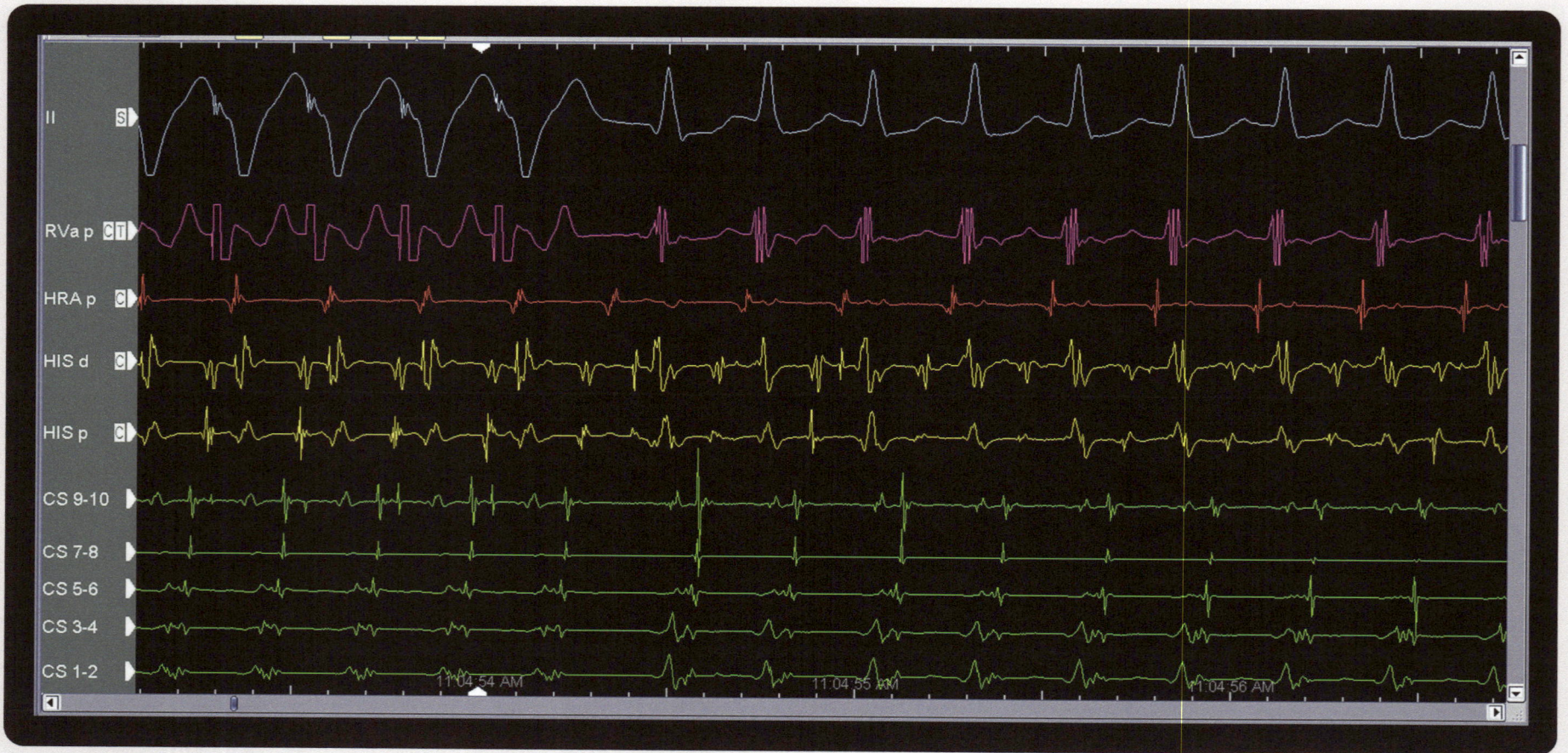

Figure 7

Figure 7: Entrainment of tachycardia 1.Entrainment of a left-sided AVRT from the RVA catheter should indicate that the RVA catheter is far from the circuit.

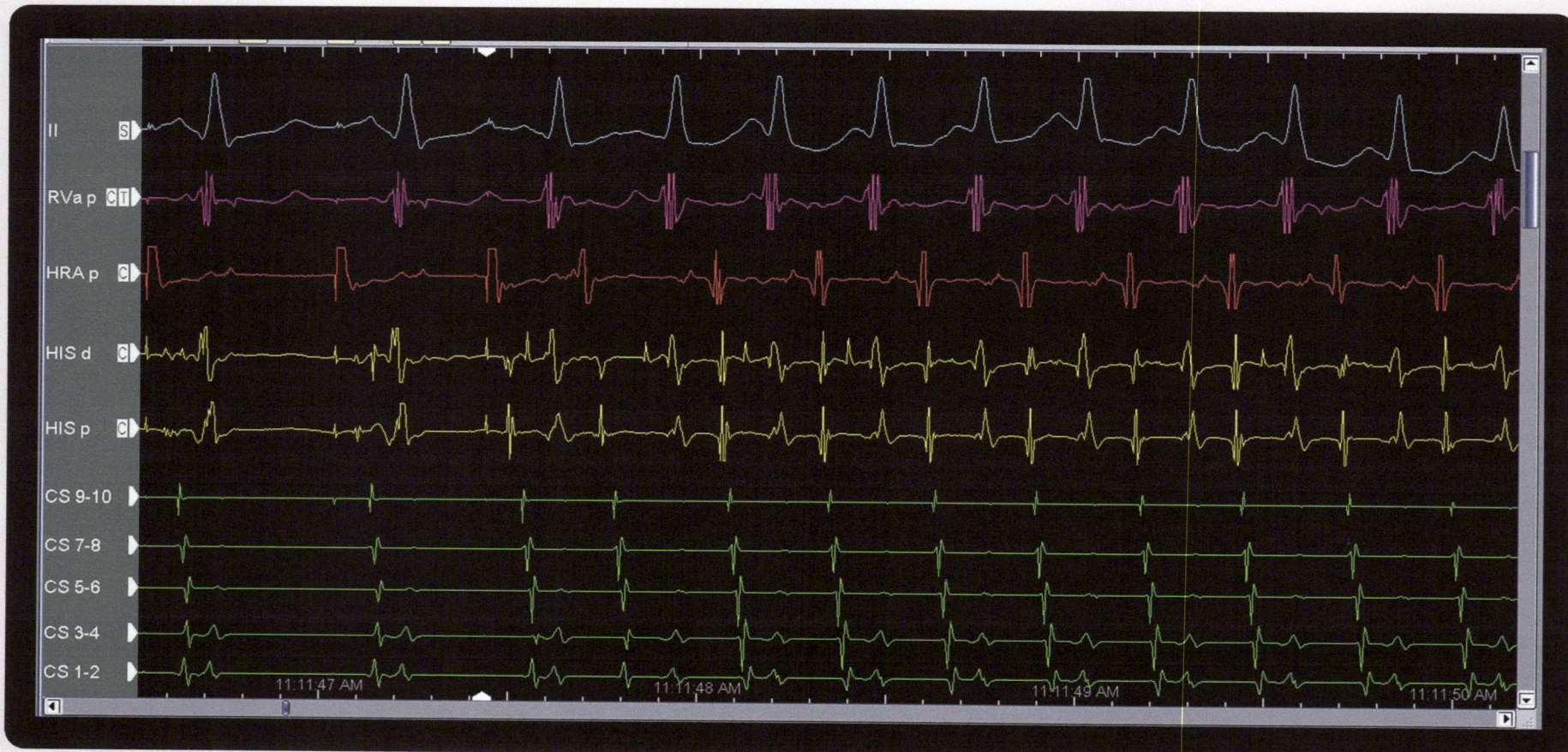

Figure 8

Figure 8: During S1S2 atrial extra stimulus pacing, tachycardia 2 is initiated. Now the HRA "A" is the earliest atrial activation. This is AVRT utilizing a right-sided AP.

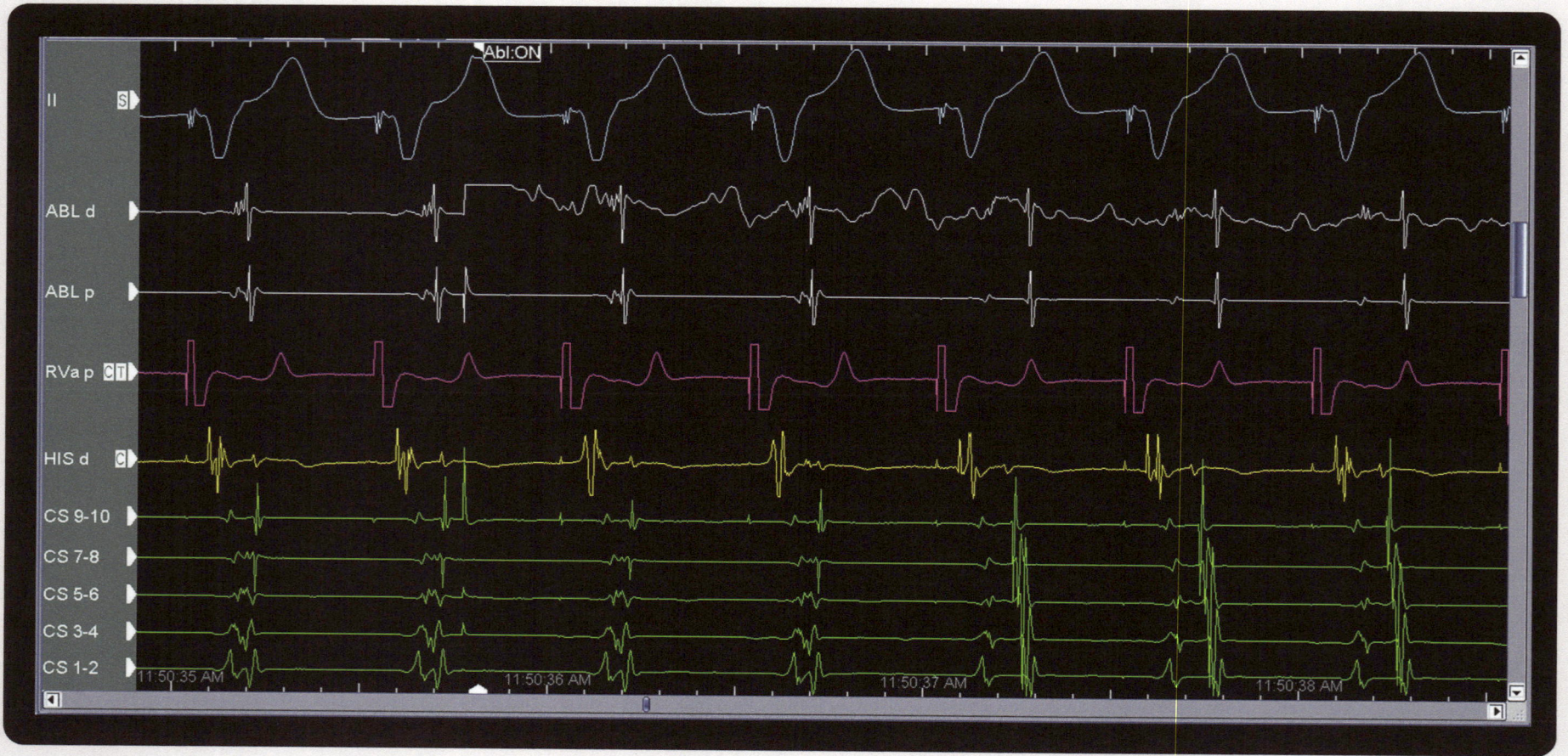

Figure 9

Figure 9: Ablation for tachycardia 1. Note the atrial activation change on the third beat after "Abl ON." The left-lateral AP has been ablated.

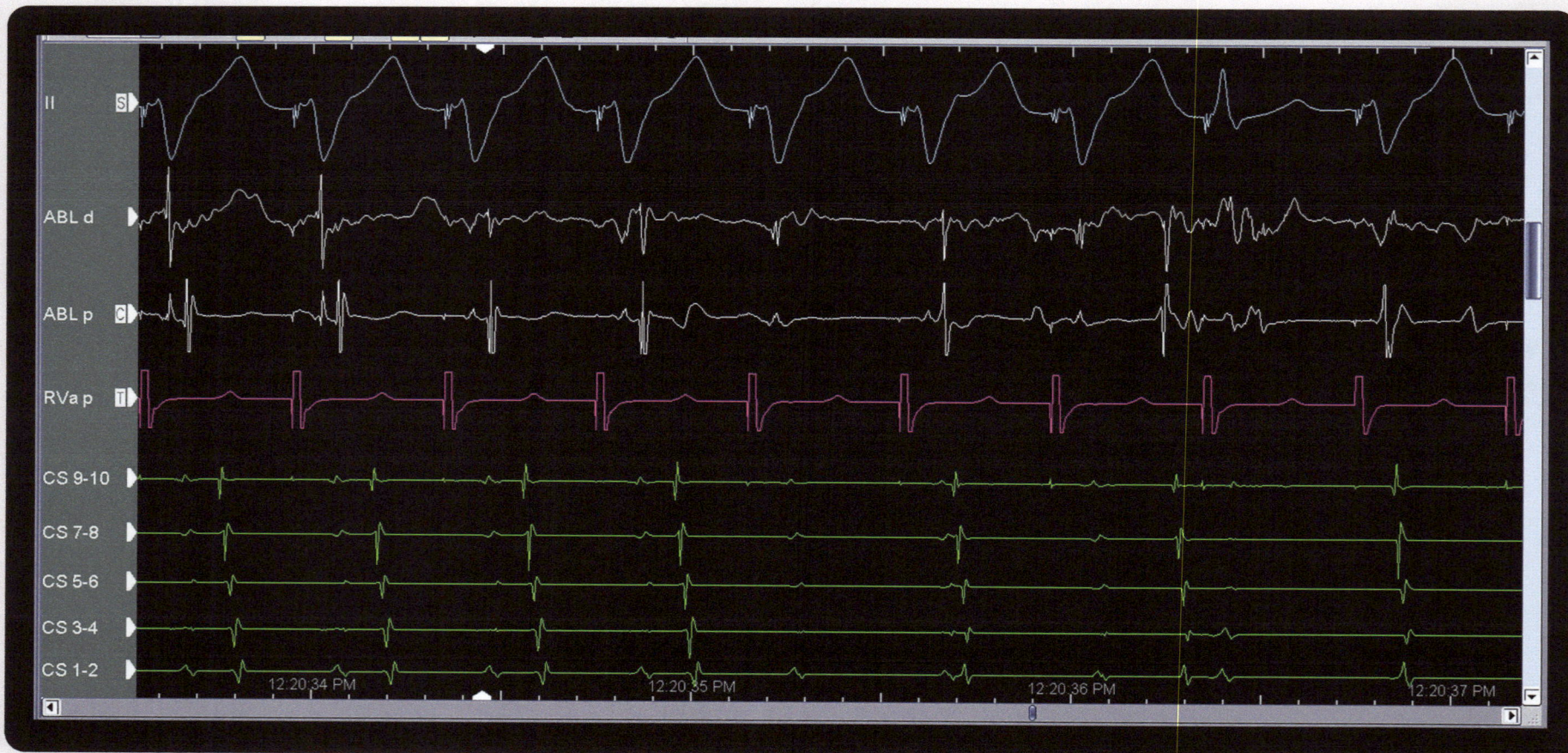

Figure 10

Figure 10: Ablation for tachycardia 2. Note the lack of V-A conduction on the fifth beat. The right-sided AP has been ablated.

This patient had a left-lateral AP *and* a right side AP!

6.2 Case 2

Here is a series of tracings presented in the order
that they were performed. Read through each tracing and
establish the diagnosis.

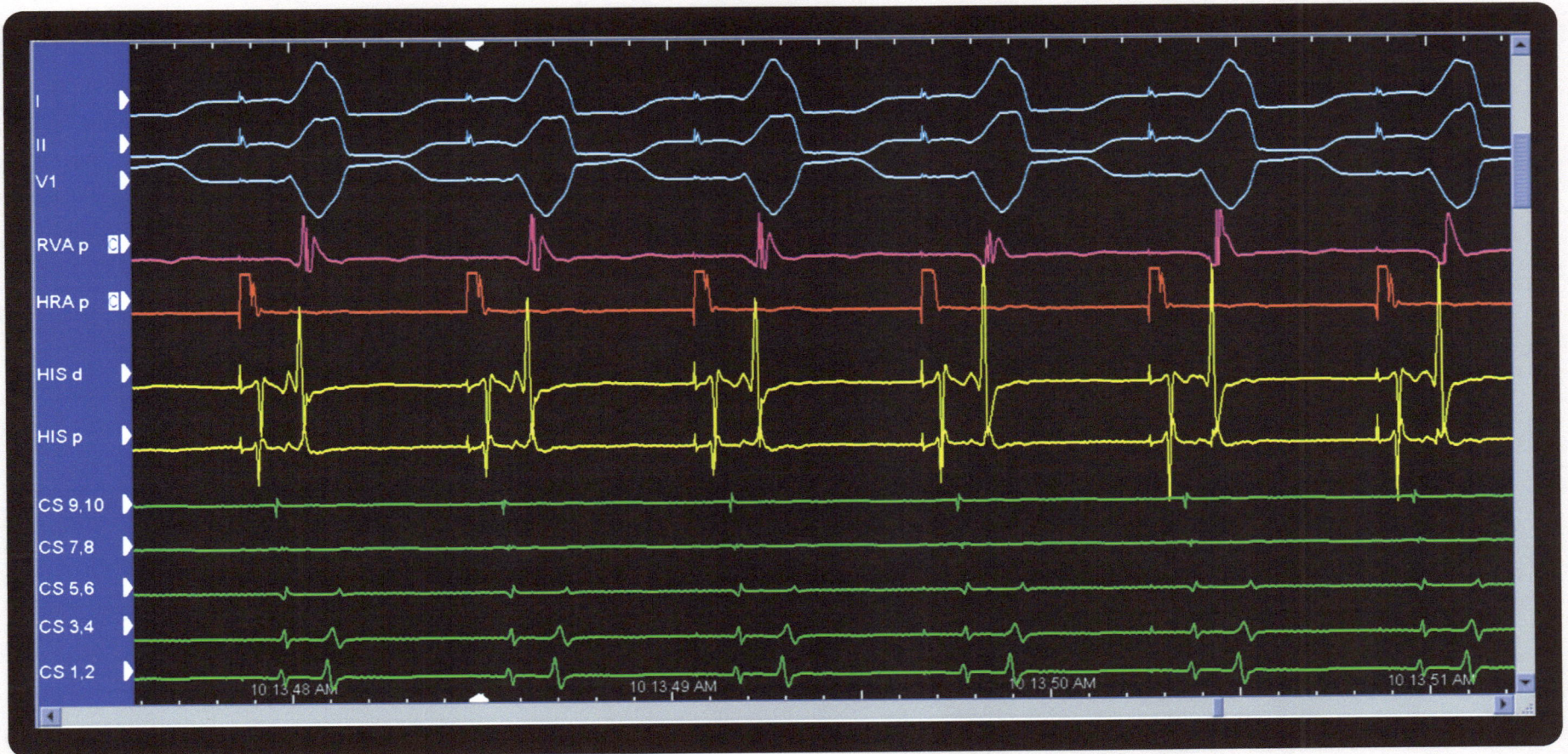

Figure 1

241

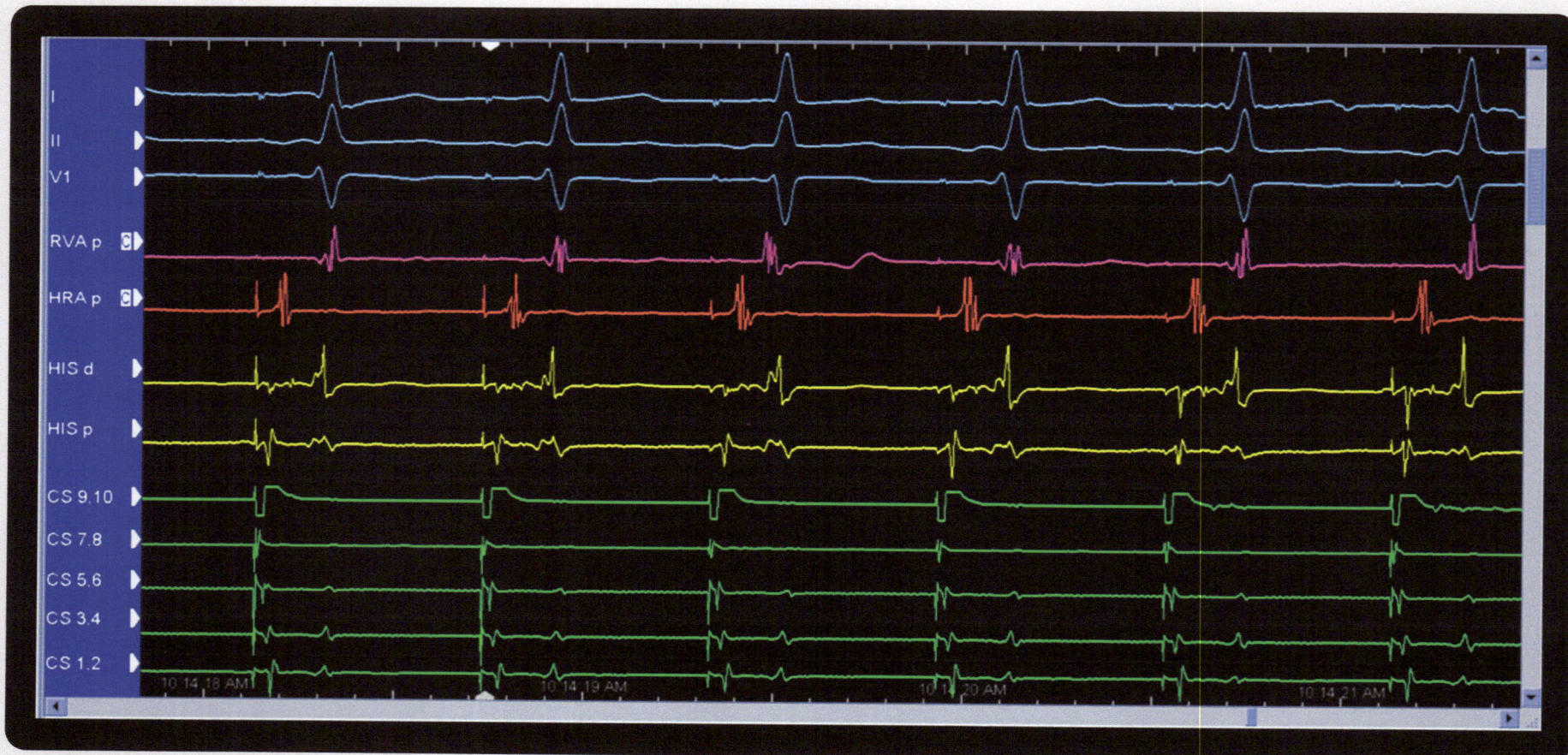

Figure 2

Discussion

Figure 1: Note that the His "V" is earlier than the RVA "V." Pacing the HRA produces a wide-complex QRS. This must be either aberrancy or preexcitation. If it was aberrancy, the His "V" would *not* be earlier than the RVA "V." So, it must be preexcitation.

Figure 2: We are now pacing CS 9-10 with a narrow-complex QRS. This occurs when the pacing site has poor access to the AP.

These first 2 tracings demonstrate that pacing different sites produce different amount of preexcitation depending on the pacing site's access to the AP.

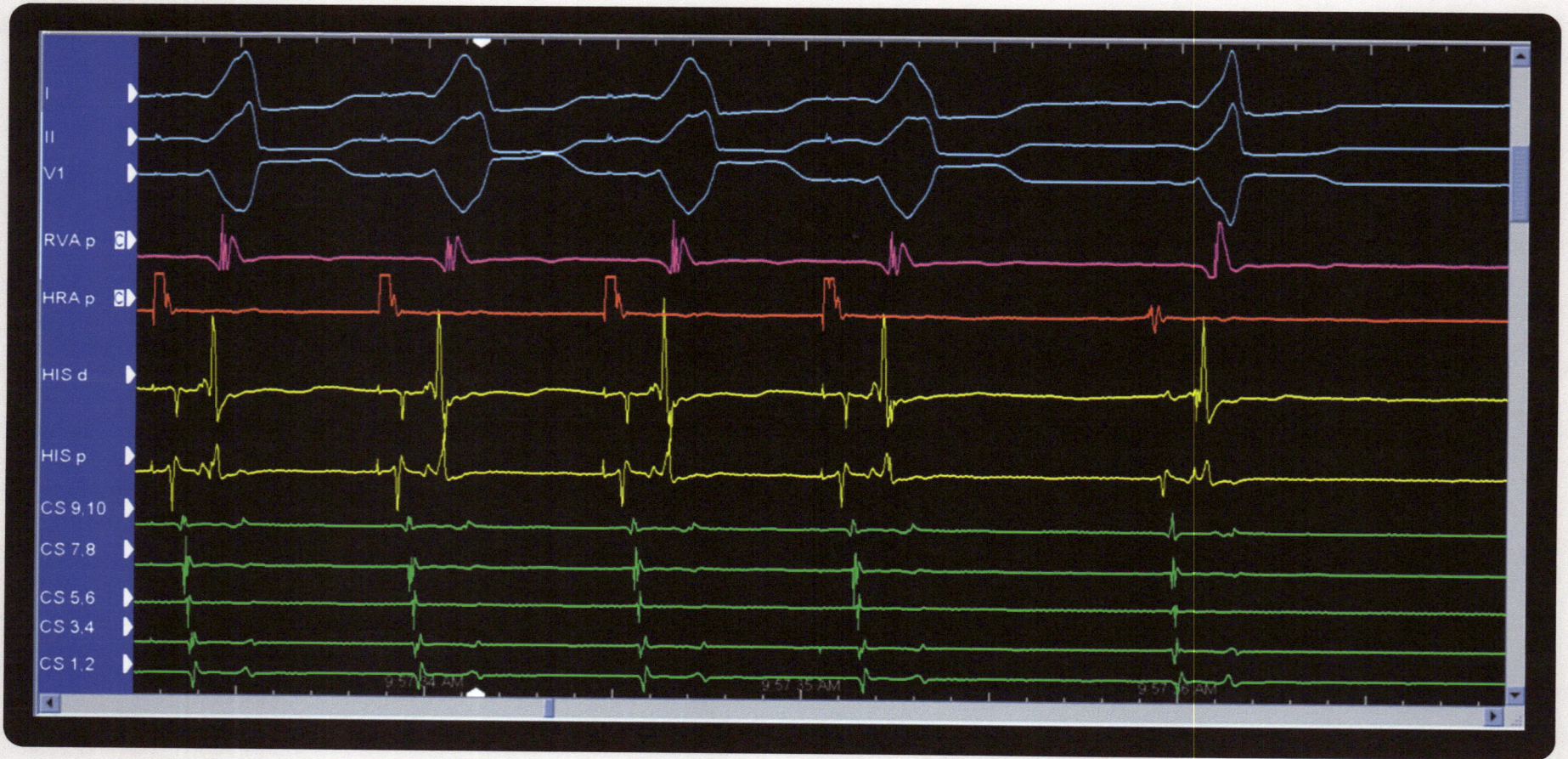

Figure 3

Figure 3: HRA extra-stimulus pacing at 600,560. The His EGM is not seen as it is hidden within the "V" electrogram.

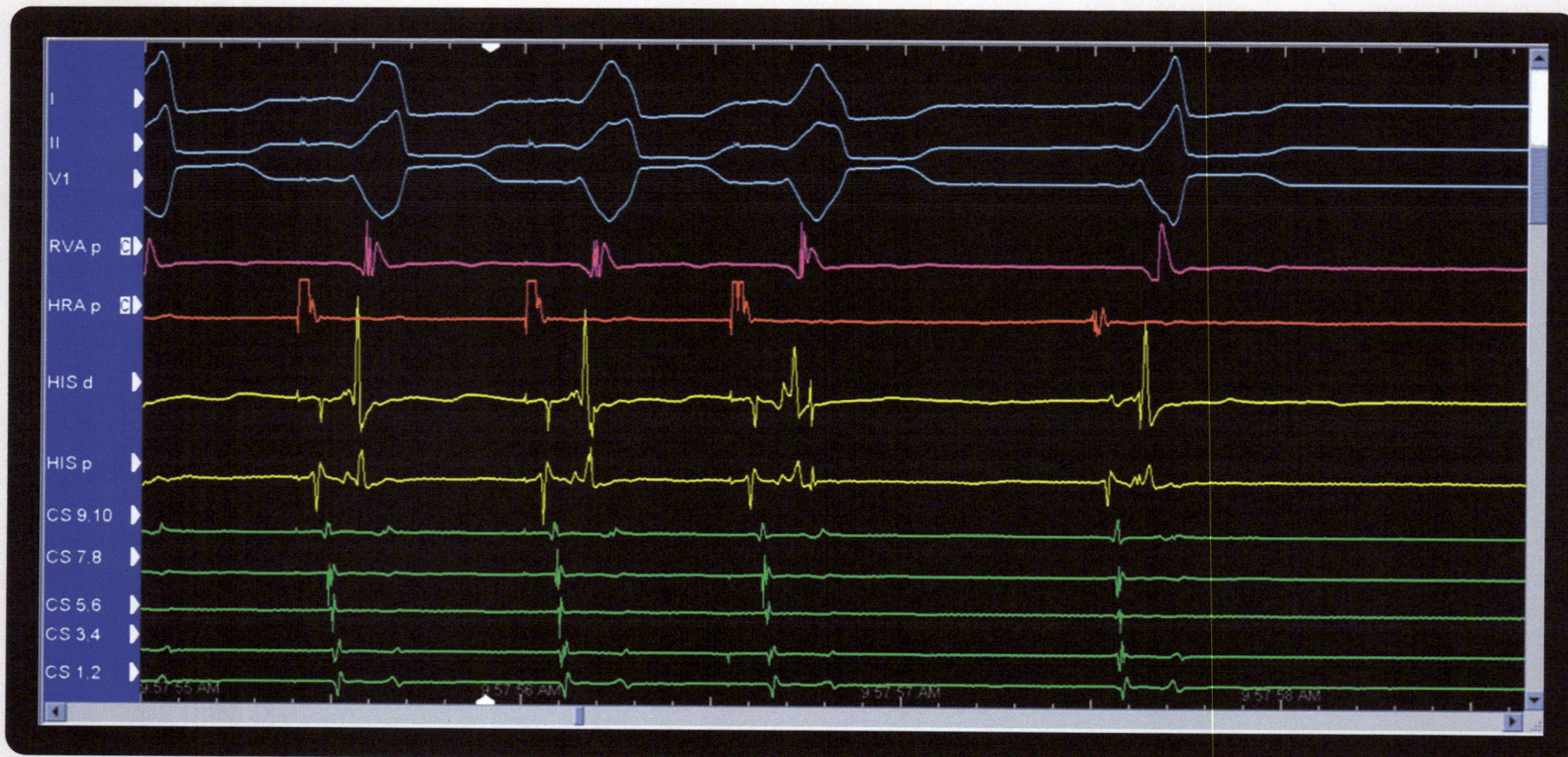

Figure 4

246

Figure 4: With a shorter S1S2 coupling interval, the His EGM is observed after the His "V" EGM!

This sequence of 2 tracings demonstrate that, in this patient, conduction has more than one route into the ventricles—through the AP and also through the AV node.

There has either been sufficient delay in the AV node that an *antegrade* His deflection appears after the His "V" electrogram, *or* this is potentially a *retrograde* His deflection. Comparison of the morphology of the His electrograms at the proximal and distal poles of the His catheter may be helpful in making this distinction.

Another possibility is, at this coupling interval, the AV node blocks in the antegrade fast pathway, and jumps to an antegrade slow pathway thus generating a late His electrogram.

6.3 Case 3

Here is a series of tracings presented in the order that they were performed. Read through each tracing and establish the diagnosis.

This is a case of an ECG suggesting one possible diagnosis and the diagnostic EP study demonstrating another.

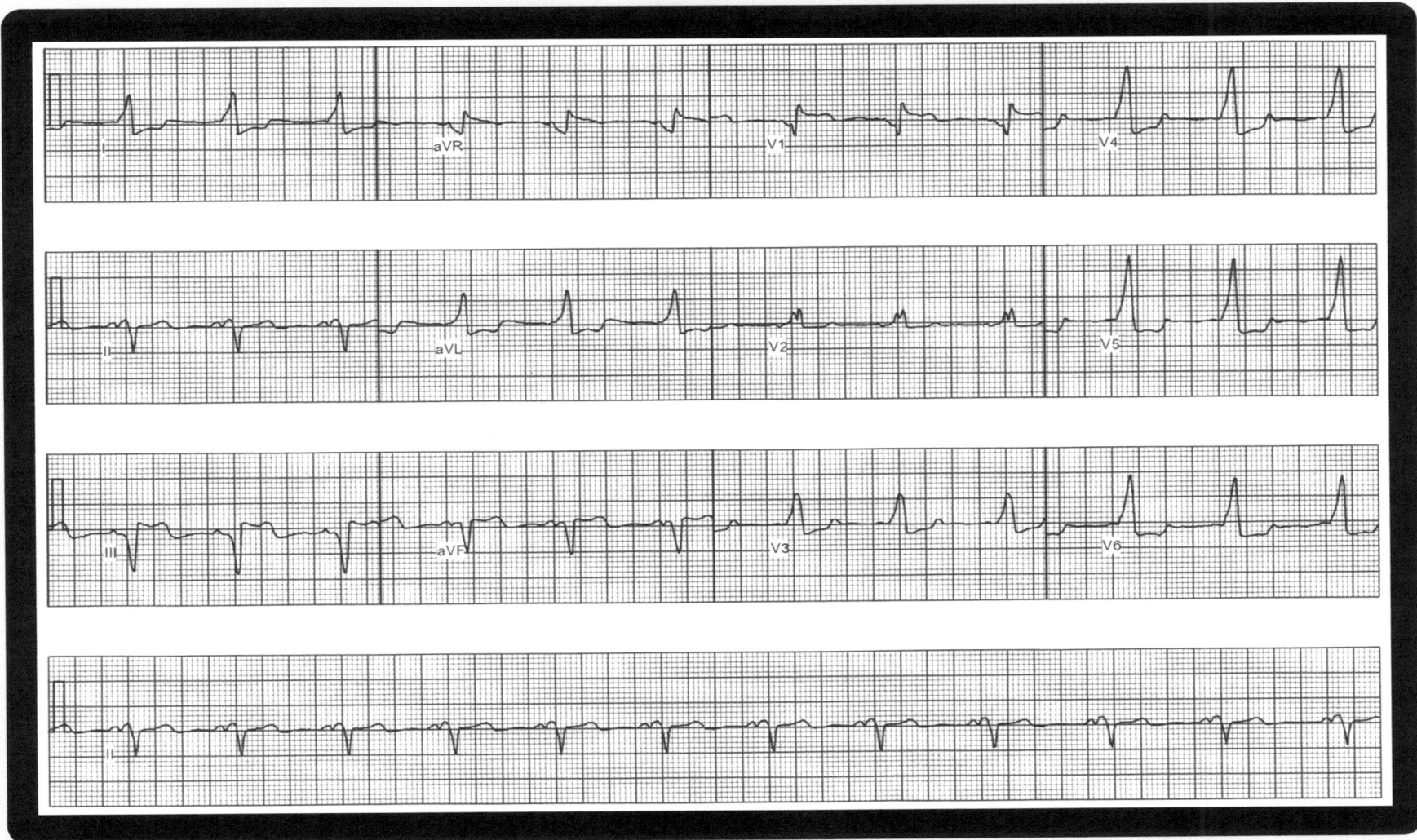

Figure 1

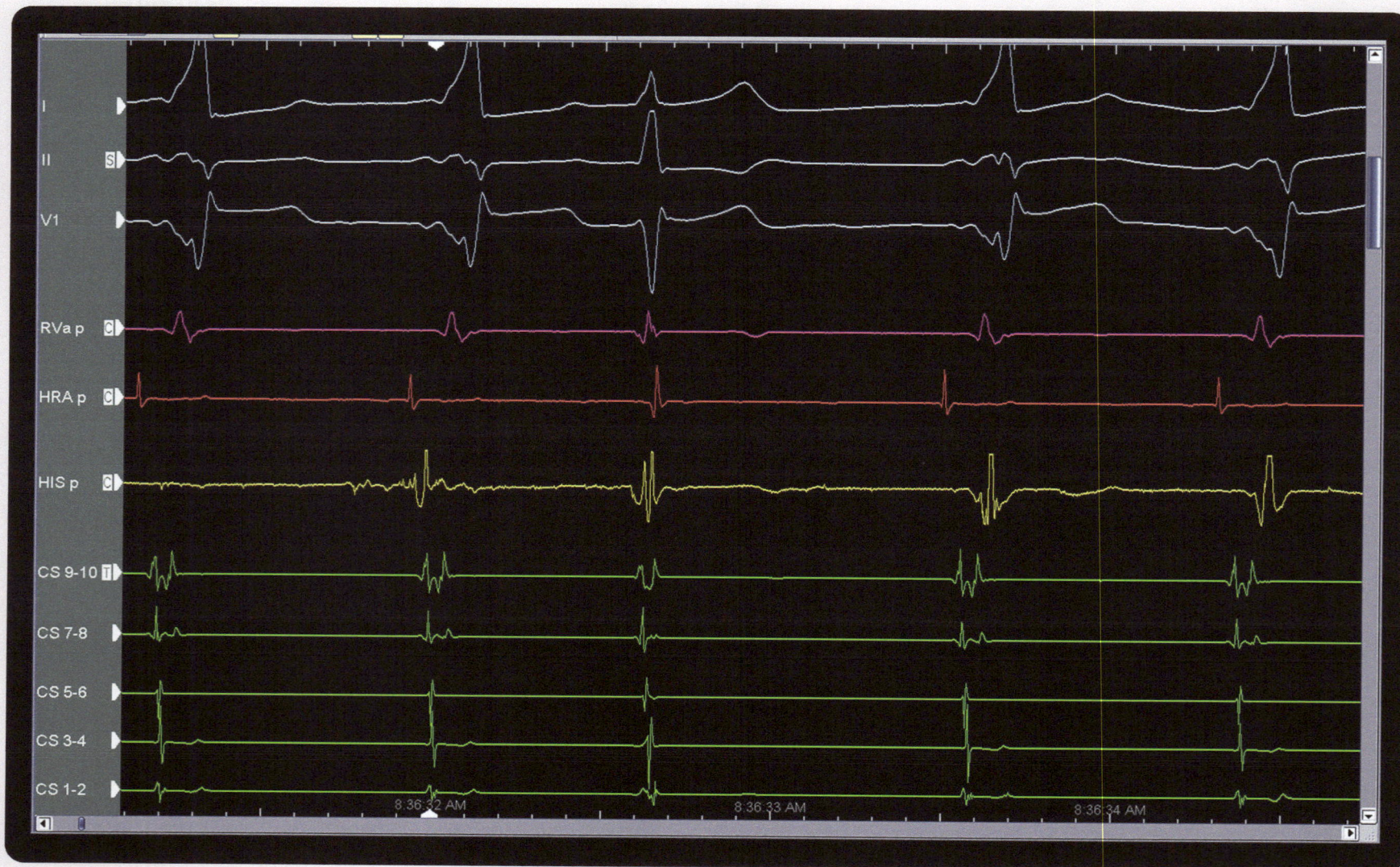

Figure 2

Discussion

Figure 1: The 12-lead ECG suggests the possibility of a posteroseptal AP.

Figure 2: While catheters are being inserted, we observed this His extrasystole. The atrial activation appears central and it is a narrow QRS.

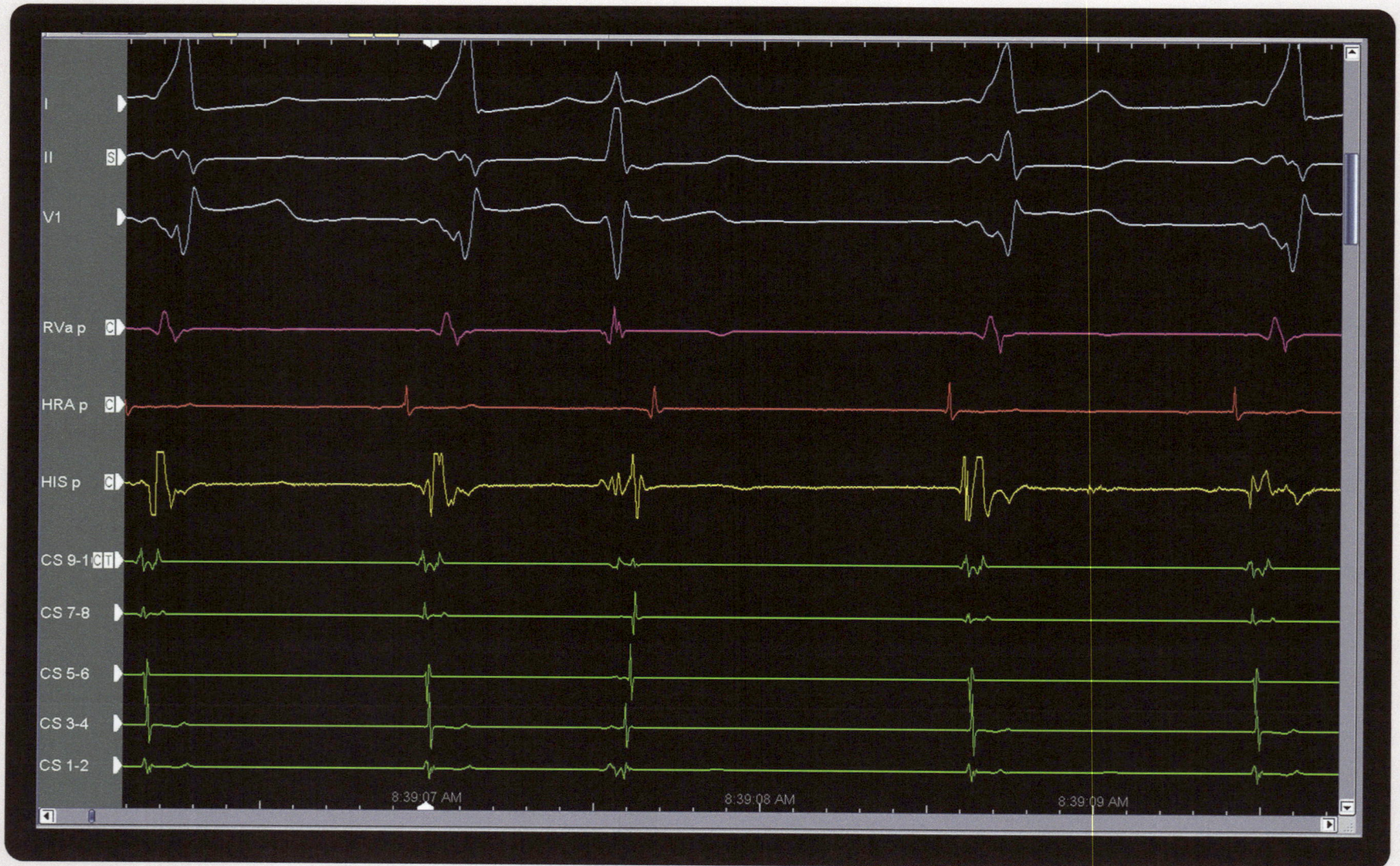

Figure 3

Figure 3: Another His extrasystole. This time the atrial activation is eccentric with the distal CS 1-2 "A" being the earliest.

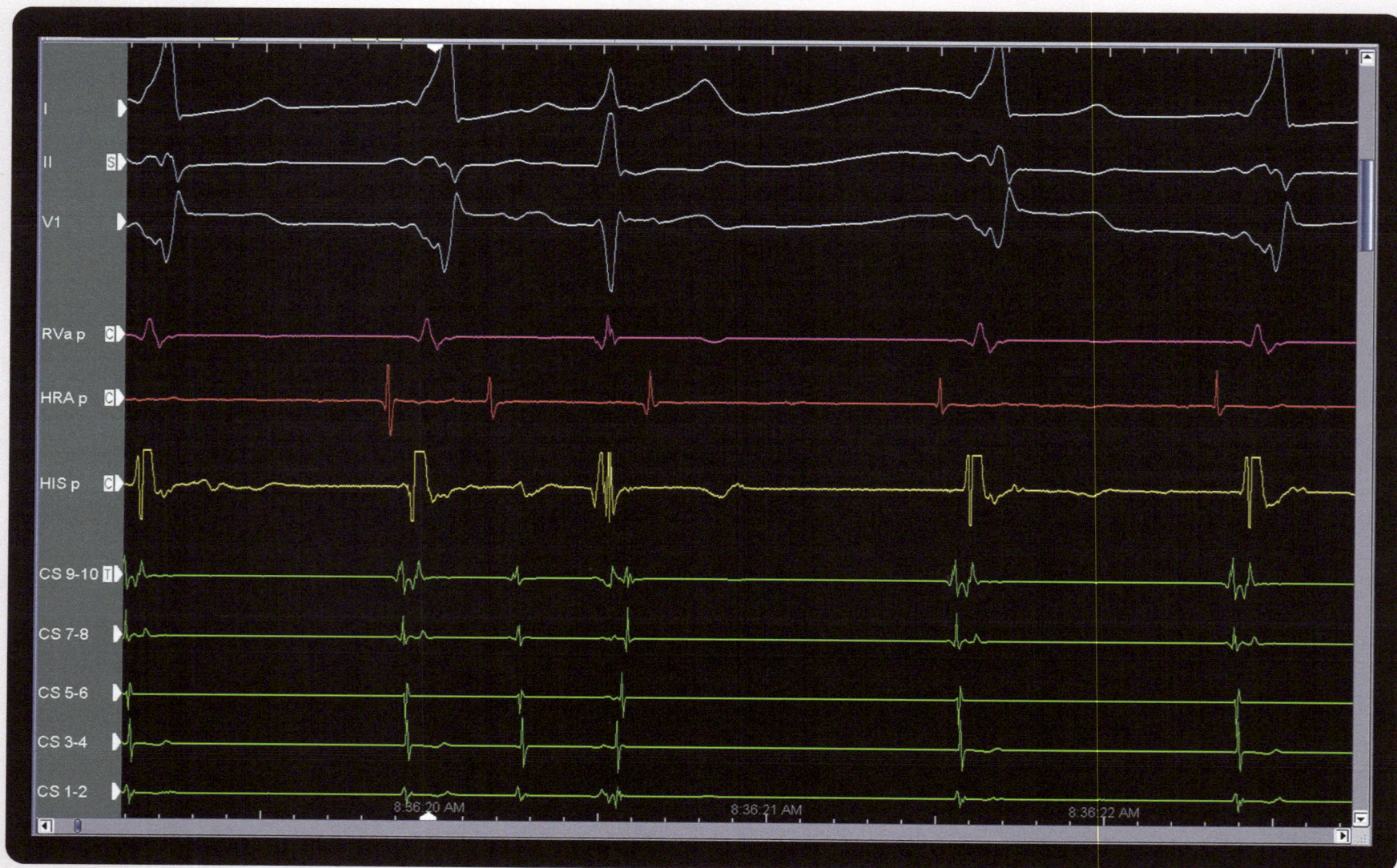

Figure 4

Figure 4: This is a PAC from the HRA catheter. It conducts
through the AV node and echoes back up the left lateral AP.

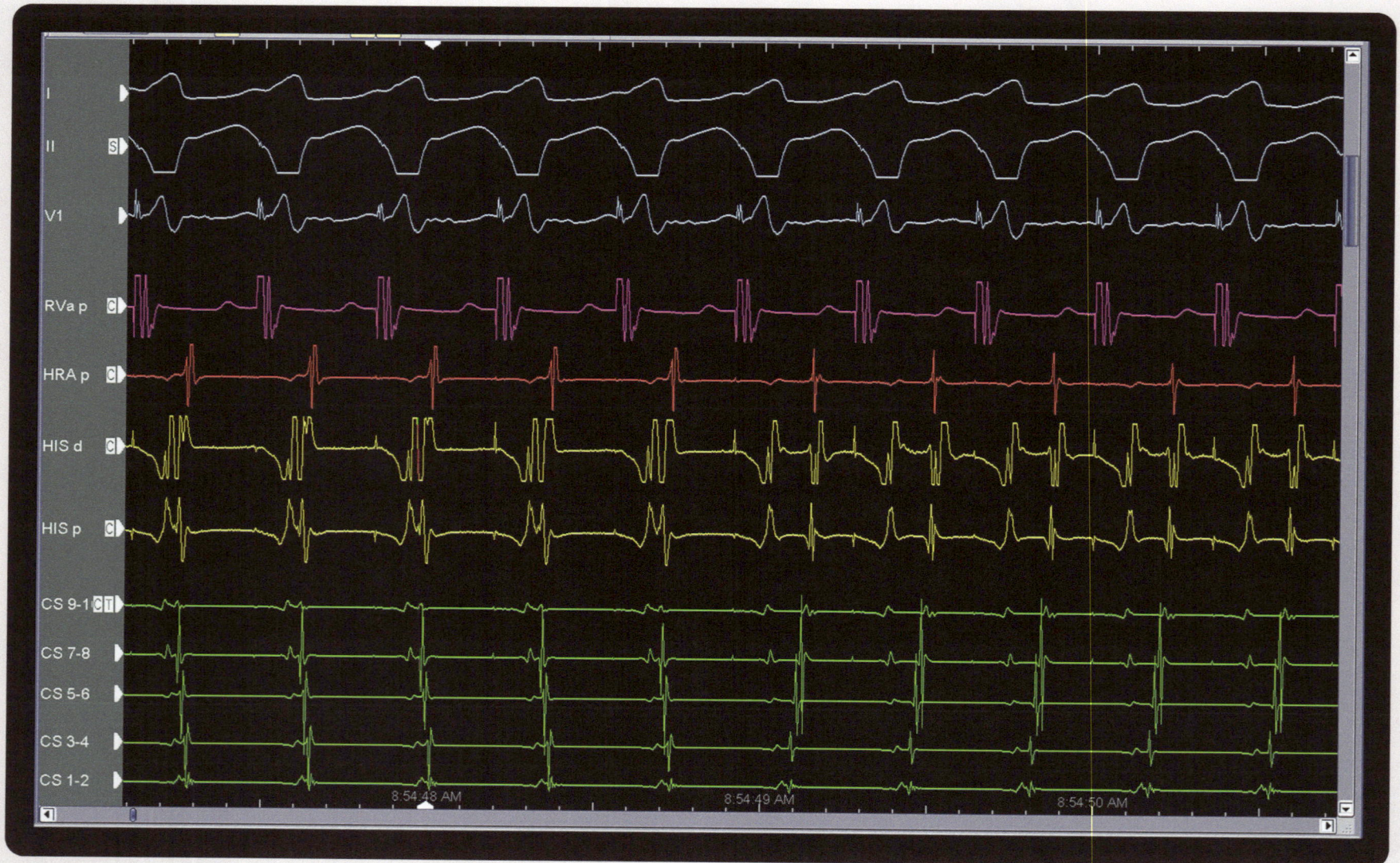

Figure 5

Figure 5: During incremental "V" pacing there is a transition from a central to an eccentric atrial activation pattern (nodal to a left lateral AP).

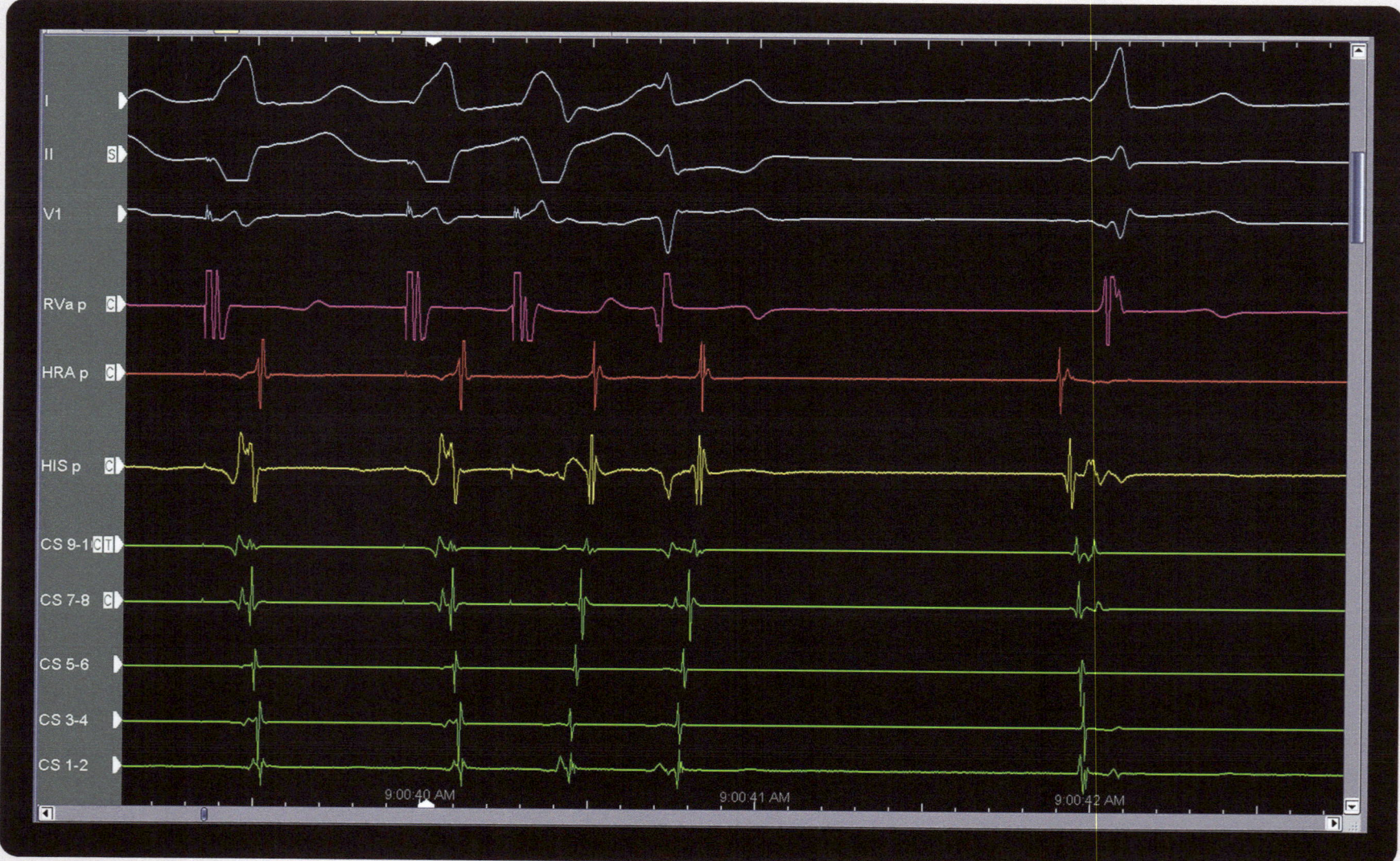

Figure 6

Figure 6: With "V" extras, the S2 blocked in the AV node. Conduction to the atria was via the left lateral AP. The narrow QRS is due to conduction down the AV node, followed by an echo up the AP.

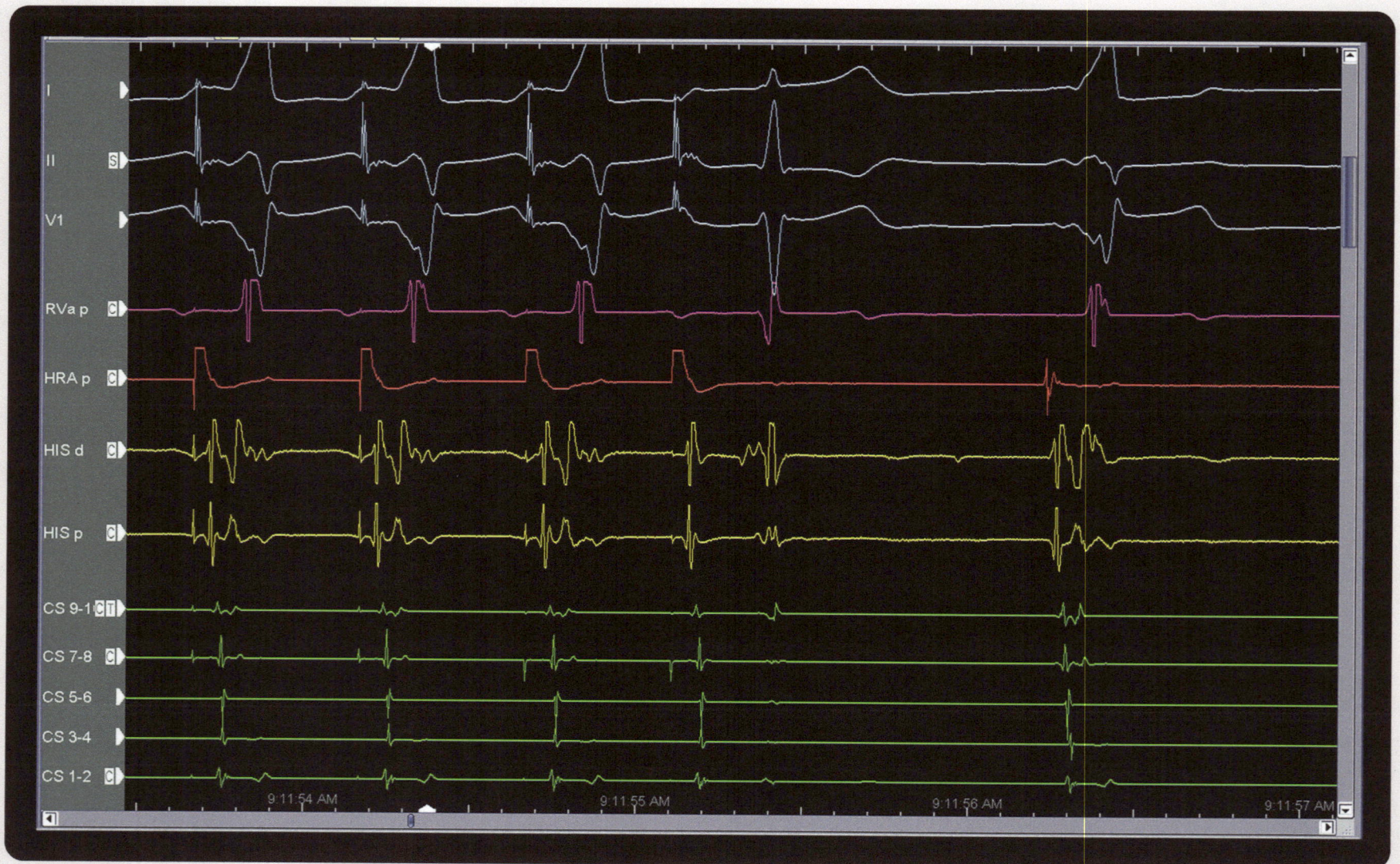

Figure 7

Figure 7: With atrial extra-stimuli, the ERP of the AP is demonstrated, resulting in a narrow QRS complex.

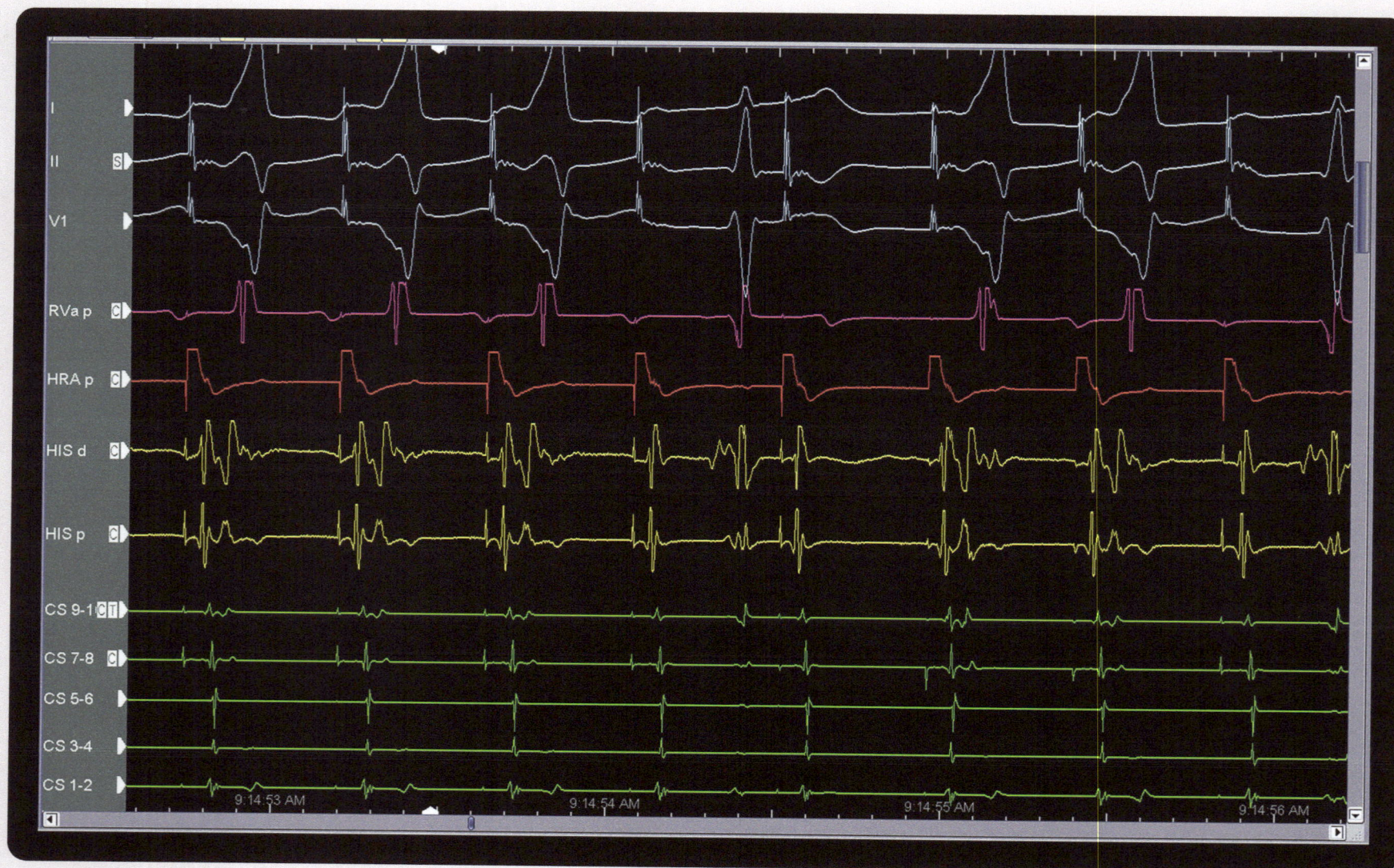

Figure 8

Figure 8: With incremental "A" pacing, we have reached the refractory periods of both the AP and the AV node.

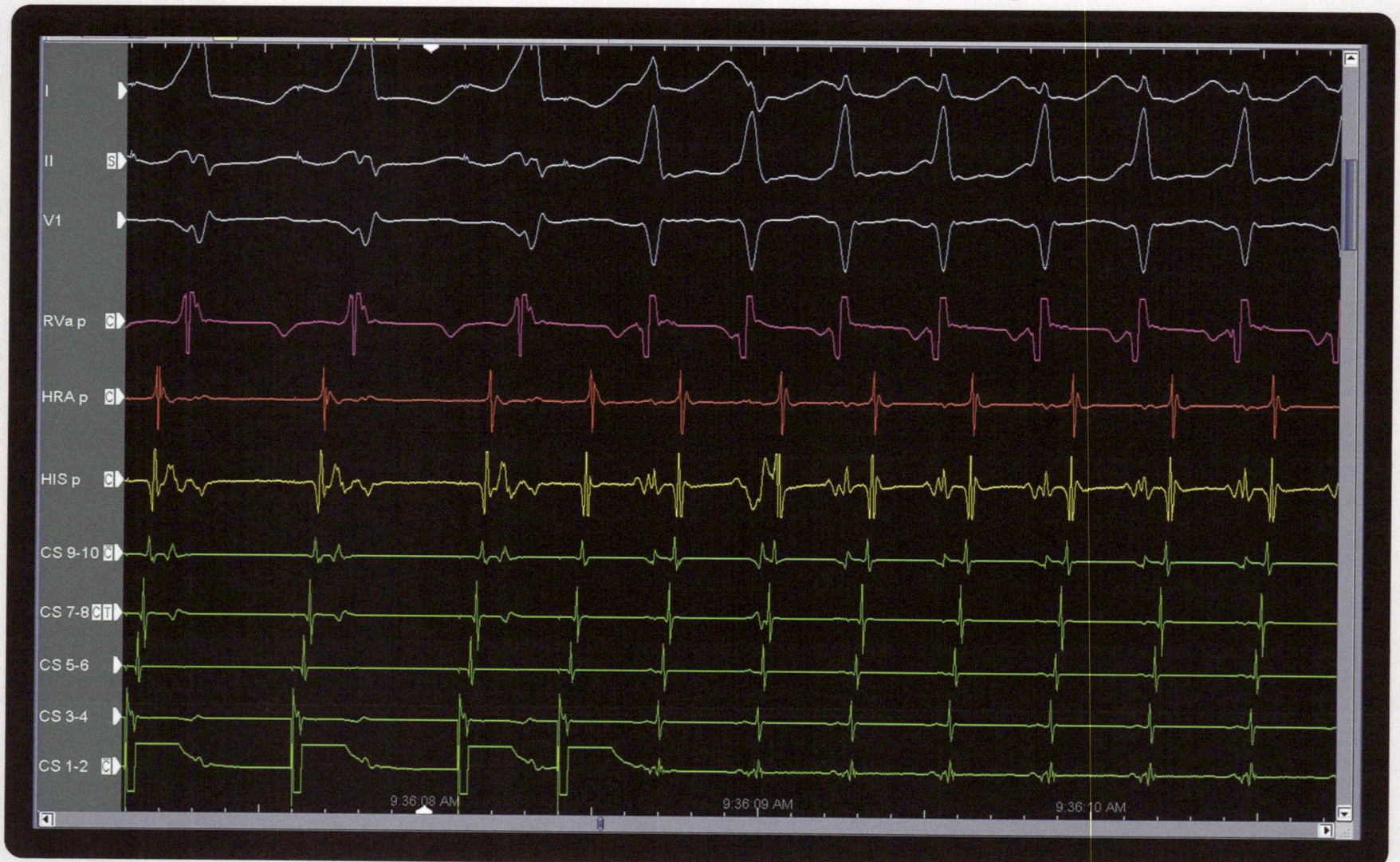

Figure 9

Figure 9: Since tachycardia has not been initiated, an alternate pacing site CS 1-2 was used. The result was induction of AVRT using a left lateral AP. By changing the pacing site relative to the location of the AP, a longer time for AP conduction recovery can lead to initiation of tachycardia.

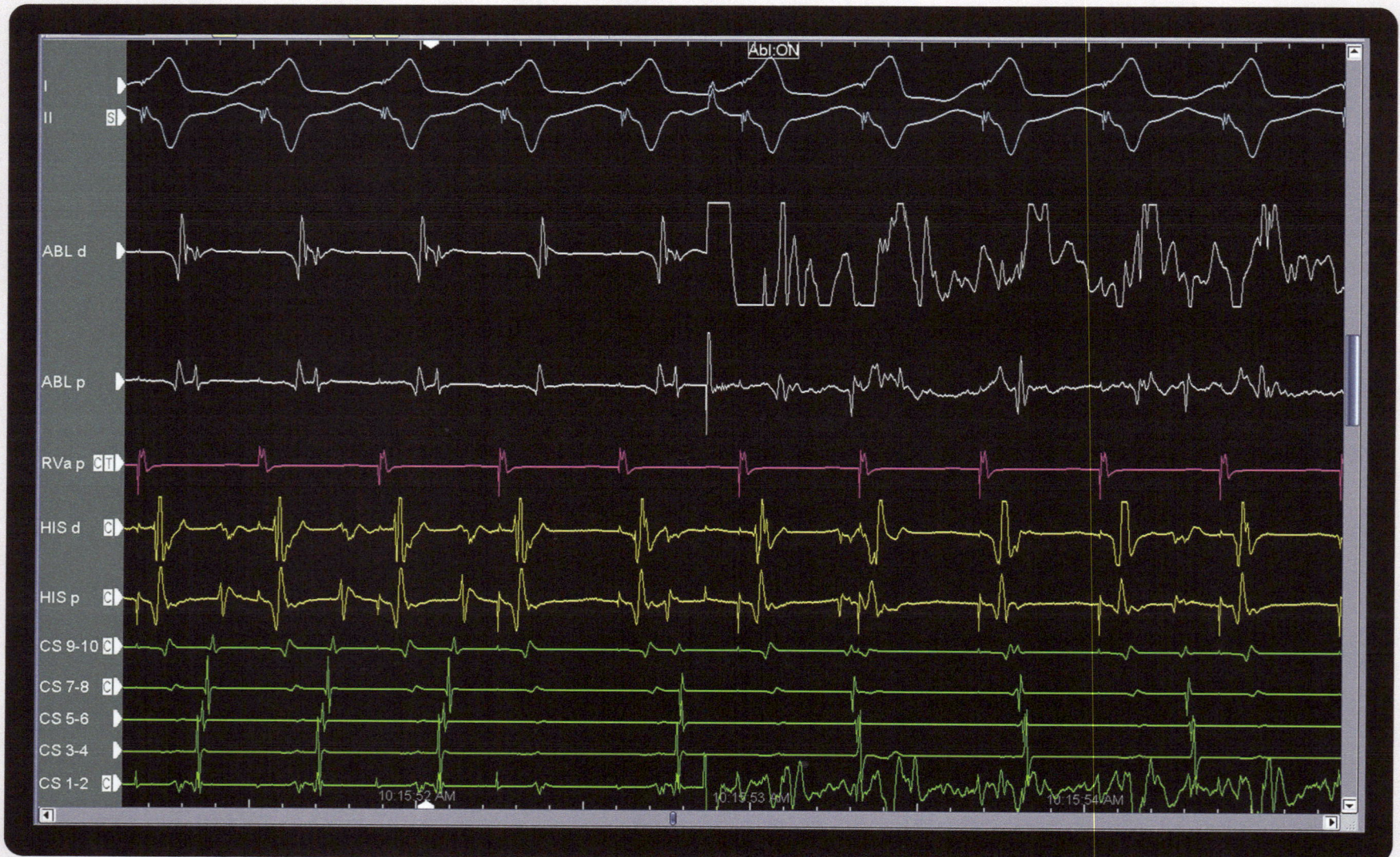

Figure 10

Figure 10: Successful ablation of the left lateral AP while "V" pacing.

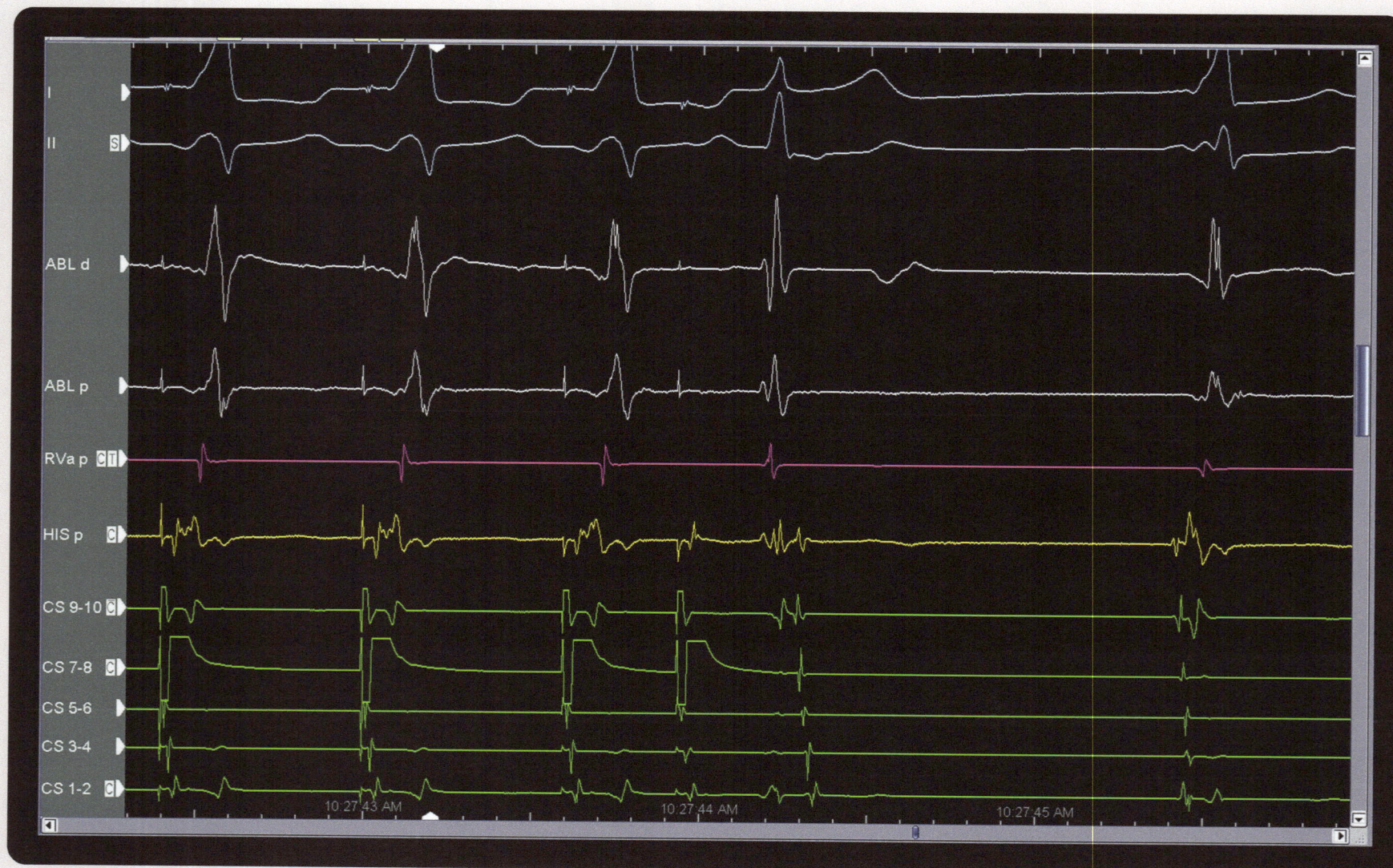

Figure 11

Figure 11: Further extra-stimulus pacing from CS 7-8 was performed because the ECG was still preexcited. The S2 reached the ERP of the AP. The S2 went down the node and then echoed back up a second, posteroseptal AP.

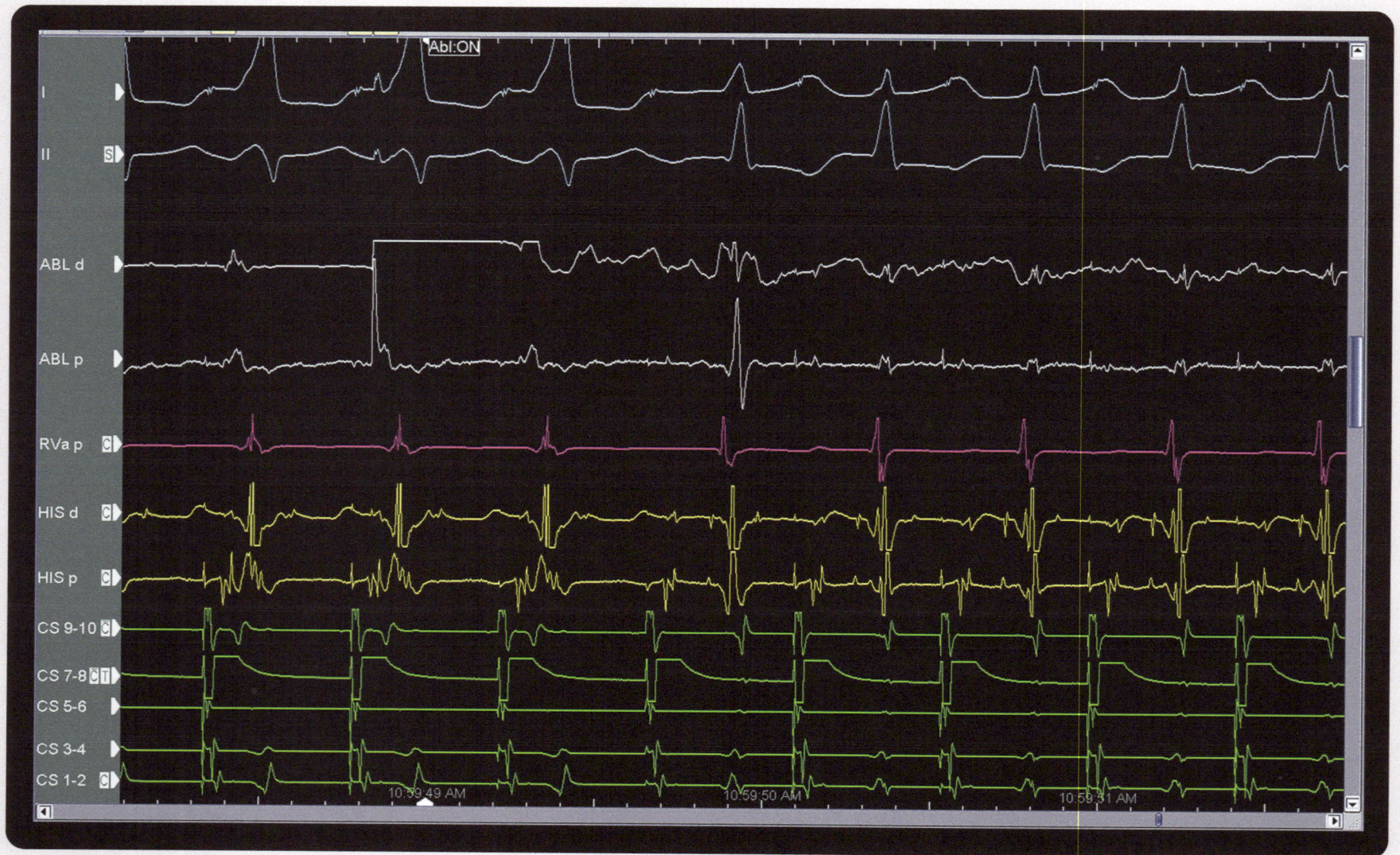

Figure 12

Figure 12: Successful ablation of the posteroseptal AP
while atrial pacing.

This is another case where the initial surface ECG suggests one
pathology (posteroseptal AP) while another (left-lateral AP)
becomes apparent on intracardiac testing.

Notes

www.ingramcontent.com/pod-product-compliance
Lightning Source LLC
Chambersburg PA
CBHW041259210326

41598CB00009B/845